Lecture Notes in Mechanical Engineering

Lecture Notes in Mechanical Engineering (LNME) publishes the latest developments in Mechanical Engineering - quickly, informally and with high quality. Original research reported in proceedings and post-proceedings represents the core of LNME. Volumes published in LNME embrace all aspects, subfields and new challenges of mechanical engineering. Topics in the series include:

- Engineering Design
- Machinery and Machine Elements
- Mechanical Structures and Stress Analysis
- Automotive Engineering
- Engine Technology
- Aerospace Technology and Astronautics
- Nanotechnology and Microengineering
- Control, Robotics, Mechatronics
- MEMS
- Theoretical and Applied Mechanics
- Dynamical Systems, Control
- Fluid Mechanics
- Engineering Thermodynamics, Heat and Mass Transfer
- Manufacturing
- Precision Engineering, Instrumentation, Measurement
- Materials Engineering
- Tribology and Surface Technology

To submit a proposal or request further information, please contact the Springer Editor of your location:

China: Dr. Mengchu Huang at mengchu.huang@springer.com
India: Priya Vyas at priya.vyas@springer.com
Rest of Asia, Australia, New Zealand: Swati Meherishi at swati.meherishi@springer.com
All other countries: Dr. Leontina Di Cecco at Leontina.dicecco@springer.com

To submit a proposal for a monograph, please check our Springer Tracts in Mechanical Engineering at http://www.springer.com/series/11693 or contact Leontina.dicecco@springer.com

Indexed by SCOPUS. The books of the series are submitted for indexing to Web of Science.

More information about this series at http://www.springer.com/series/11236

Lucas F. M. da Silva · Robert D. Adams ·
Chiaki Sato · Klaus Dilger
Editors

Industrial Applications of Adhesives

1st International Conference on Industrial Applications of Adhesives

Springer

Editors
Lucas F. M. da Silva
Department of Mechanical Engineering
Faculdade de Engenharia
University of Porto
Porto, Portugal

Robert D. Adams
Department of Mechanical Engineering
University of Bristol
Bristol, UK

Department of Engineering Science
University of Oxford
Oxford, UK

Chiaki Sato
Materials and Structures Laboratory (MSL)
Tokyo Institute of Technology
Yokohama, Japan

Klaus Dilger
Institut für Füge- und Schweißtechnik
Technische Universität Braunschweig
Braunschweig, Niedersachsen, Germany

ISSN 2195-4356 ISSN 2195-4364 (electronic)
Lecture Notes in Mechanical Engineering
ISBN 978-981-15-6766-7 ISBN 978-981-15-6767-4 (eBook)
https://doi.org/10.1007/978-981-15-6767-4

This Springer imprint is published by the registered company Springer Nature Singapore Pte Ltd.
The registered company address is: 152 Beach Road, #21-01/04 Gateway East, Singapore 189721, Singapore

Contents

Use of UV-Curing Adhesive Systems on Non-transparent Joining Parts by Using Sidelight Activated Polymer Optical Fibres

R. Seewald, J. Kallweit, J. Weiland, A. Schiebahn, T. Gries, and U. Reisgen

Abstract The long curing times of the adhesives are an essential deficit of the bonding technology. UV-curing adhesives enable complete curing within seconds, but require at least one light-transparent joining partner. In order to make use of the advantages of UV-curing adhesives also on non-transparent joining parts, the approach of light transfer into the adhesive joint is pursued. A polymer optical fibre (POF) is embedded in the adhesive as a waveguide. With a lateral activation of the fibre surface, a radial curing is realized. In this paper, the effects of side light activation by thread cuts and gradual micro cuts on a PMMA POF with a diameter of 1.0 mm are characterized and validated in corresponding adhesive tests. An experimental setup is established to characterize the activated fibre with regard to sidelight intensity and homogeneity. The wavelengths of the UV light used are 365 and 400 nm. While lateral homogeneous adhesive layer depths of up to 2.5 mm are achieved at 400 nm, the use of UV light at 365 nm proves to be unsuitable for the selected tests.

Keywords Polymer optical fibres (POF) · Sidelight characterization · UV curing adhesive

1 Introduction

UV adhesives are used in large numbers for precise positioning and alignment of components. These joints can be used for structural or temporary bonding. UV curable adhesives can provide good adhesion to many substrates including metals, rubber, glass, ceramics and plastics such as polycarbonates, acrylics and phenolic resins. Therefore, UV adhesives are used extensively in packaging, electronics,

R. Seewald (✉) · J. Weiland · A. Schiebahn · U. Reisgen
ISF—Welding and Joining Institute, RWTH Aachen University, Pontstraße 49, 52062 Aachen, Germany
e-mail: seewald@isf.rwth-aachen.de

J. Kallweit · T. Gries
ITA—Institut für Textiltechnik, RWTH Aachen University, Otto-Blumenthal-Straße 1, 52074 Aachen, Germany

Lucas F. M. da Silva et al. (eds.), *Industrial Applications of Adhesives*, Lecture Notes in Mechanical Engineering,
https://doi.org/10.1007/978-981-15-6767-4_1

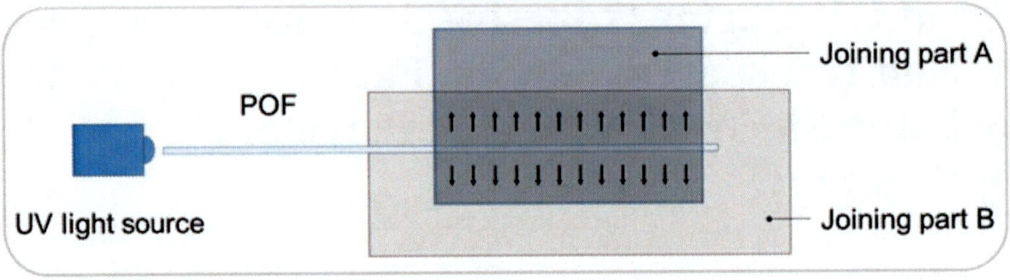

Fig. 1 Concept of FibreKleb technology

medical, transport, industrial assembly and glass bonding. Properties such as optical clarity and transparency, dimensional stability, low shrinkage and resistance to chemicals and solvents are among the advantages of UV adhesives [1]. A major advantage of UV adhesives over other types of adhesives widely used in industry is the short curing time. For example, while 2-component epoxy resin adhesives usually have curing times of 15 min to 7 days, UV adhesives can be completely cured within seconds.

Up to now, the use of UV adhesives has only been possible with at least one UV-light transparent part. The use of UV adhesives on non-transparent parts is not possible according to the current state of the art. The specific advantages of UV adhesives, such as curing within a few seconds, simple dosage by only one adhesive component and the almost infinite open time have so far not been able to be used for joining parts that are not transparent for UV light.

With the use of optical polymer fibres (POF) it becomes possible to cure UV adhesives between non-transparent UV light materials. A POF is inserted into the bonded joint, via which UV light is coupled out radially or axially to cure the adhesive in the bonded joint. Since the POF consists of a polymer, the mechanical properties of the POF can be specifically adapted to the bond. This process called FibreKleb is being researched by the institutes ISF and ITA of RWTH Aachen University. A schematic illustration is shown in Fig. 1.

An external UV light source induces UV light into the POF. This emits light laterally in the area of the bond and starts the curing reaction of the UV adhesive.

Polymer optical fibres (POF) are waveguides which are based on the principle of total internal reflexion (TIR). In terms of carrying light intensity from one point to another any refraction is to be avoided. This means that, speaking in terms of geometric optics, light rays that do not exceed the critical angle of total internal reflection when reaching the fibre surface are reflected theoretically without any intensity loss. Thus, in order to make POF emit light laterally through their surface, TIR is to be avoided and refraction as a way of partial reflection and transmission is desirable. This can be done in different manners as can be seen in Fig. 2 [2].

The sidelight activation process for POF needs to be adapted to the objectives of a sidelight homogeneity and intensity. Due to internal absorption effects and the emitted light intensity, the sidelight activation needs to gradually increase for a homogeneous laterally emitted light. This is why the process of surface roughening

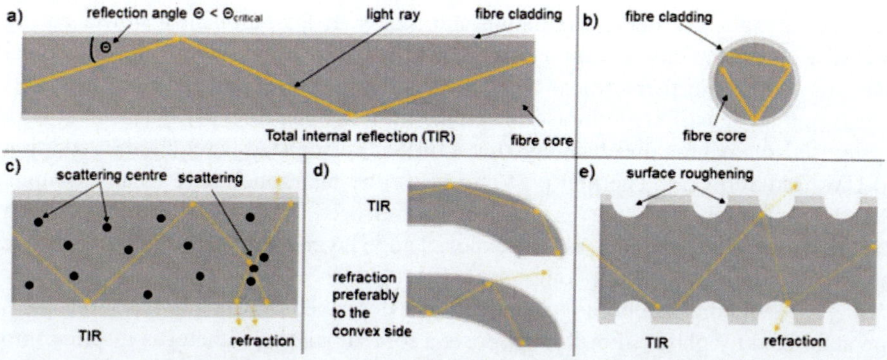

Fig. 2 Polymer optical fibres as optical waveguides. **a, b** Total internal reflection (TIR) in an optical fibre in fibre axis and cross-sectional view, **c–e** sidelight activation by incorporation of scattering particles, bending and surface treatment

is the preferred sidelight activation option for POF available on the market and in this study as well.

In this article, the FibreKleb technology is investigated on POF made of polymethyl methacrylate (PMMA). Polymethyl methacrylate is a transparent thermoplastic with excellent light transmission and very good light and media resistance. For this reason, the plastic is used in numerous optical, electronic and lighting applications. Different pre-treatment methods for the lateral activation of the fibre are considered to enhance the lateral output of UV light. A commercially available sidelight fibre is used as a reference. The fibres are characterized in terms of intensity and homogeneity of lateral light emission. A uniform method for characterization will be established in order to achieve a high reproducibility and comparability of the different experimental parameters. The results of the fibre characterization are then validated in adhesive tests and process limits are determined. The aim of the research is the substitution of existing slow bonding processes by a faster UV process and thus a reduction of high cycle times. This enables new fields of application to be developed across all sectors.

2 State of the Art

At the current state of the art, conventional UV adhesives can only be used with at least one transparent substrate, since UV adhesives only cure where UV-light exposure occurs. Therefore, no curing takes place in shadow areas. Until now, the problem of shadow curing has been solved by chemical modification of the adhesive. For this purpose, cationic reaction mechanisms or other curing functions such as thermal curing are integrated into the adhesive (so-called dual cure systems) [3]. While radical UV adhesives cure directly under the influence of light, cationic systems activate the adhesive under irradiation and thus start a significantly slower curing process. However, this has the consequence that the short curing time of radical systems is lost. Further critical aspects of cationically reacting UV-curing adhesives are their sensitivity to moisture and the limited open time [4]. These dual-cure systems also have the disadvantage that a further curing step is required in addition to UV irradiation [3]. Thermal post-curing can be mentioned here as an example. Another approach is the design of specially mirrored irradiation chambers, which can, however, also irradiate only the exposed edge layers of adhesive joints and have no effect on non-transparent components [5].

In [6, 7] a medical technology application is described in which broken bone parts are stabilized by plastic. For this purpose, a folded balloon catheter is inserted into the bony canal and pushed inside the bone cavity until behind the fracture. In the next step, the balloon catheter is filled with a UV light-curing liquid methacrylate (MA) plastic. This expands the balloon catheter and supports itself on the walls of the inner bone. The MA plastic is cured by a glass fibre in the catheter, through which UV light is introduced for curing. According to [8] the MA plastic is cured in 6–10 min. According to [7], the maximum curing depth radially to the glass fibre is 11 mm. It is not further specified which wavelength of UV light is used for curing the MA plastic. The papers [6, 7] also do not go into further detail as to whether adhesive bonding forces can form at interfaces—in this sense, it is not a joining process. In addition, the behaviour of the interface between glass fibre and MA plastic, especially under the influence of mechanical stress, is not further characterised or described in the papers.

In [9, 10] the application of POF and glass fibres in in-mould processes is described. The authors call the development a "Closed Mould Photocuring System". A cationic curing epoxy resin is used as UV-curing resin. As POF a PMMA-POF with a Polyvinylidene fluoride (PVDF) cladding with a diameter of 1.2 mm is considered. PVDF is a chemically quasi-inert polymer, which therefore does not allow any chemical/physical bonding to the matrix. Different variants for the improvement of the lateral light radiation are investigated by various treatments, such as permanent modification of the fibre geometry with adjusted bend radii as well as by mechanically embedding silica scattering particles into the fibre and applying micro-cuts. However, no experiments on the applicability of the technology for a joining process are performed here.

Glass optical fibre (GOF) have previously been used for curing of UV adhesives. The choice of glass fibres is justified by the lower intensity loss for UV light in GOF compared to POF [11]. However, from a mechanical-technological point of view, glass fibres prove to be unsuitable for the use in adhesive joints, as their typical modulus of elasticity is about 70,000 N/mm^2 whereas common UV adhesive systems have moduli of elasticity between 175 and 3000 N/mm^2. In contrast, POF (made of PMMA) have a typical modulus of elasticity of 3200 N/mm^2 and therefore behave similar to the adhesive joint under mechanical deformation (especially under load parallel to the grain direction). The latter means that the POF introduced into the bonded joint is not a disturbing element.

An expired patent claims the joining of components, in particular car body parts, by means of a UV adhesive with GOF is reported. The GOF has previously been damaged specifically to form coupling points for the UV laser radiation [12]. As explained before, GOF are mechanically less suitable for the UV adhesives than POF. Furthermore, no further publications or market launch of [12] could be found.

3 Materials and Methods

Three different fibre systems were used for the following adhesion tests. Their properties are shown in Table 1.

The aim was to investigate a standard POF without side light activation available on the market (BK-40) and also a POF that is already available on the market with gradual side light activation (M-fibre). It should be noted that the M-fibre actually is a diffuser tube with four individual fibres and only one single fibre of the tube was investigated. In addition, fibres were spun out on a monofilament melt spinning line with the structure shown schematically in Fig. 3. The spinning parameters stretching and water bath temperature were optimized in a factor test plan. The filament (VT7) with the lowest achieved optical attenuation of 2.6 dB/m at a wavelength of 650 nm was further investigated. It was produced with a stretch ratio of 2.5 at a water bath temperature of 65 °C.

Two different UV adhesive systems are investigated in this paper. The general properties and information can be found in Table 2. The two adhesive systems were selected because of their large curing depth, since the aim of the test is to achieve the deepest possible radial curing.

Since the curing depth is dependent on the radiation intensity and the irradiation time, the information given by the manufacturer is only a relative comparison and must be tested in appropriate adhesive tests.

In order for the POF to emit UV light radially, a side light activation is required. Two different methods of surface treatment are investigated in this paper. On the one hand, defined micro-cuts are radially introduced into the fibre surface with a depth

Table 1 Characteristics of the POF under investigation

Name	BK-40	M-Fibre	VT7
Supplier	Mitsubishi Chemical Corporation, Tokyo, Japan	MENTOR GmbH & Co. Präzisions-Bauteile KG, Erkrath, Germany	Produced in-house
Core material	PMMA	PMMA	Plexiglas Optical POQ66—PMMA
Cladding material	Fluorinated polymer	Fluorinated polymer	None
Diameter [mm]	1.0	1.0	1.0
Sidelight activation	None	Abrasive	None

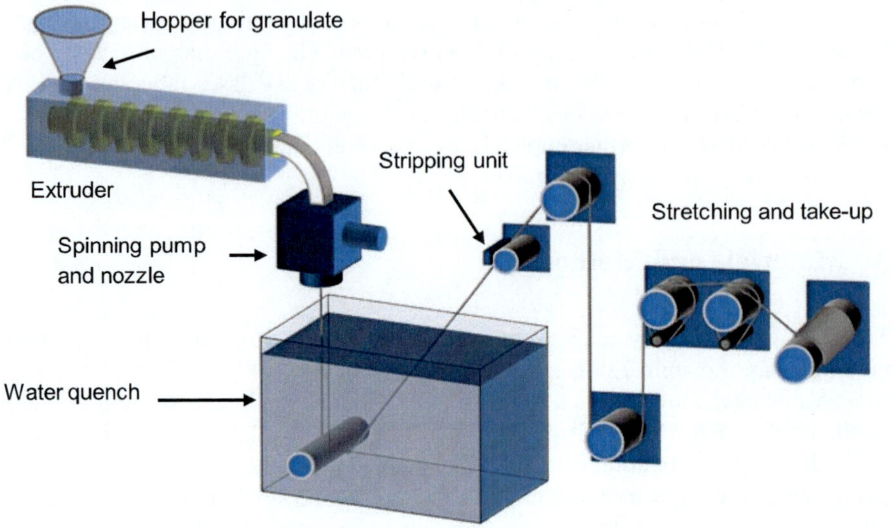

Fig. 3 Schematic diagram of a monofilament melt spinning process with a water quench [13]

Table 2 Properties of the investigated UV adhesives

Product name	Photobond FB4175	Loctite AA3525
Manufacturer	DELO Industrie Klebstoffe GmbH & Co. KGaA, Windach, Germany	Henkel AG & Co. KGaA, Düsseldorf, Germany
Chemical base	Modified acrylate	Modified acrylate
Recommended wavelength	400 nm	365 nm
Curing depth	15 mm (@ 50 mW/cm^2, 60 s)	12 mm (@ 50 mW/cm^2, 60 s)
Viscosity	210,000 mPa s	15,000 mPa s

of 15–20 μm. The distances between the cuts are gradually increasing in order to achieve a homogeneous side light emission. On the other hand, an M1.2 thread is turned on the fibre to obtain a uniform surface structure with a depth of 30–40 μm. In comparison, commercial sidelight fibres are used as a reference. The different surfaces are shown enlarged in Fig. 4. While at 10× magnification on the M-Fibre no pre-treatment of the surface can be seen, at 20× magnification the traces of a pre-treatment by particle beam become clear.

In order to characterize the effect of the surface treatments, intensity measurements are made at the fibre end for balancing purposes. For uniform and reproducible test conditions, the fibres are first cut to a length of 200 mm. The fibre ends are flattened by wet grinding with a 5000 grit sandpaper. Then a standardized connector (F-SMA905) is mounted on one fibre end to ensure reproducible coupling into the test setup. The DELOLUX50 UV LED spot lamp of DELO Industrie Klebstoffe is used

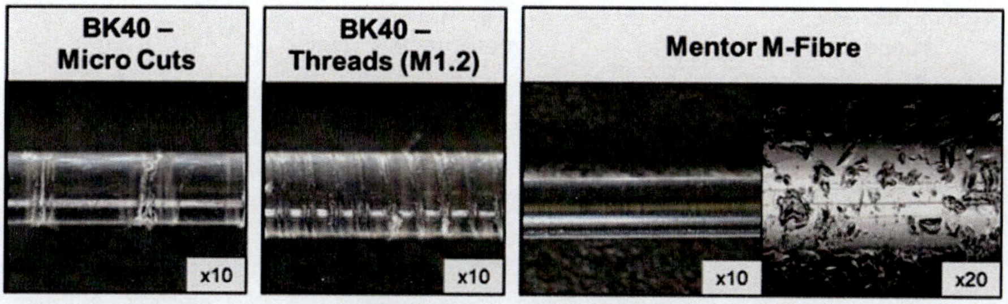

Fig. 4 Different treated surfaces of PMMA fibres with diameter 1.0 mm

as light source. The lens at the lamp head of the emitter focuses UV light with a wavelength of 400 nm to a diameter of approx. 1.1 mm and has a peak intensity of 12,000 mW/cm^2. Since the working distance and collinearity of the lamp and the test fibre have a great influence on the coupling of the UV light into the POF, a 3-axis cage system of opto-mechanics is used for positioning.

The open fibre end is placed on the detector surface of the UV intensity meter used. The device is a surface sensor for radiation measurement in the UV-A range from *Dr. Hönle AG, Gräfelfing, Germany*. Since the fibre diameter only covers a part of the sensor surface, the absolute measured values are not representative. However, a relative comparison of the measured values is appropriate due to the constant test conditions. In order to record the intensity of the laterally emitted UV light and thus to quantitatively evaluate the surface treatment, a balance is drawn between the measured intensities at the fibre end of a treated and an untreated fibre. This cancels out internal absorption effects of the fibres.

Furthermore, the emission spectrum of the laterally emitted light is recorded at three points over the test area of 50 mm using a USB2000+ spectrometer from *Ocean Insight Inc., Winter Park, Florida, USA*. The distance between measuring head and fibre is 35 mm. Between the measuring points, there is a distance of 20 mm each. Here again, the absolute measured values are not quantitatively meaningful, but a relative consideration to evaluate the lateral intensity distribution with regard to its homogeneity is possible and reasonable.

A test fibre and the division into the different coupling or measuring ranges as well as the test setup is shown in Fig. 5.

4 Characterization of the Sidelight Activation

The characterization of POF is divided into two measurements. On the one hand, the relative axial intensity loss of activated fibres compared to an untreated fibre is determined, Fig. 6. On the other hand, the lateral light spectrum is determined and normalized at three measuring points in the activated fibre area, Fig. 7.

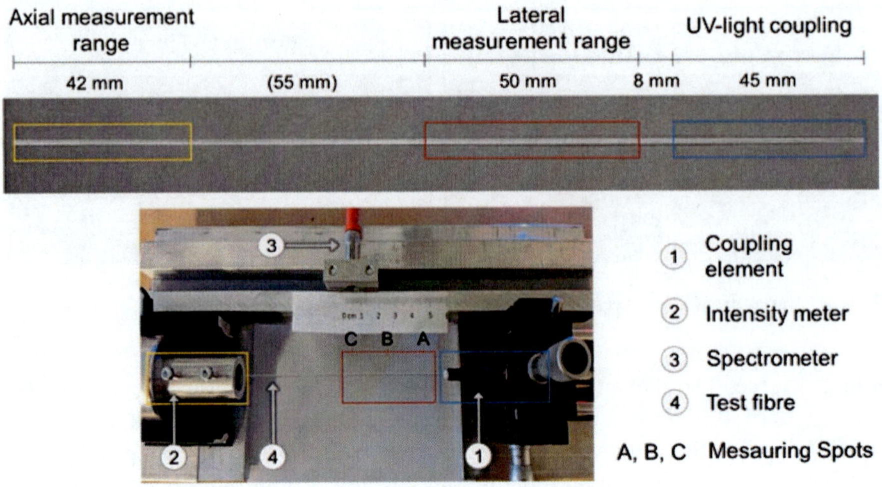

Fig. 5 Test fibre areas and test setup for POF characterization

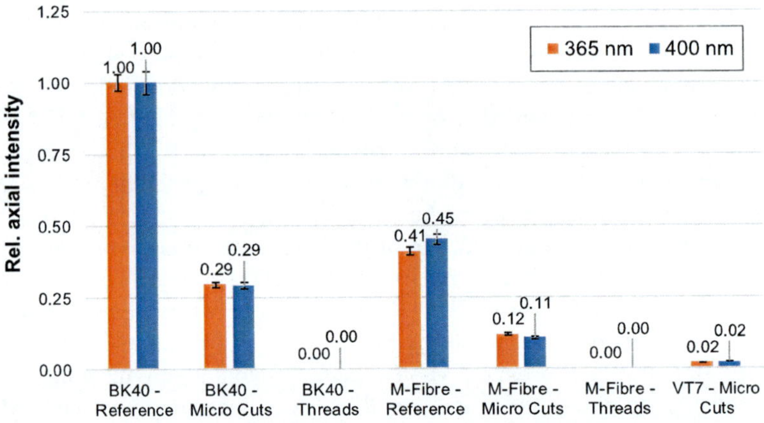

Fig. 6 Correlation between relative axial UV-intensity and different POF and surface treatments

The axial intensity measurements were carried out for both 365 and 400 nm, as these are the recommended wavelengths for the adhesives under investigation. However, no significant difference in the light wavelength is detectable, so that there is no wavelength dependence between the fibres tested. The influence of the activation methods, on the other hand, is significant. The difference between the untreated BK40 fibre and one with micro cuts is 0.71, which means that 71% of the induced light is emitted laterally. On fibres with a thread cut, even no axial intensity is measured anymore, so that almost the entire light must be emitted laterally. The commercial sidelight fibre in its original state already has an axial intensity that is almost 60% lower than the inactivated reference, which can also be further reduced to almost

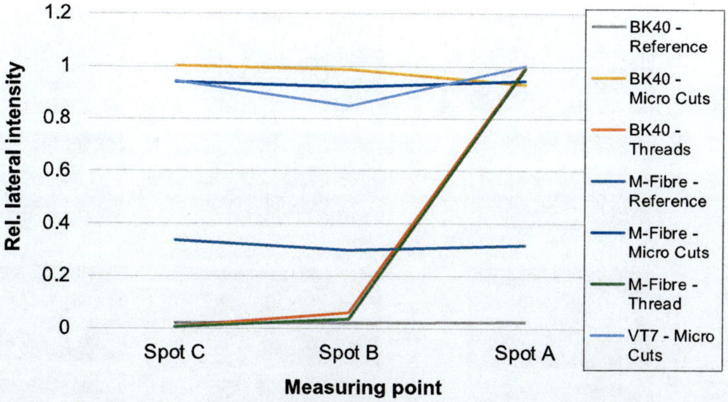

Fig. 7 Relative lateral intensity at three measuring points on different POF and surface treatments

90% under surface treatment. The VT7 fibre with micro cuts has the lowest axial intensity with over 99% intensity loss. However, the internal attenuation of the fibre is significantly higher than in commercial POF, so that a comparison between the fibres is only conditionally valid.

The measurement and evaluation of the laterally measured intensities gives a conclusion on the homogeneity of the fibre activation. The highest intensities are found at the beginning of the activated area (Spot A) for fibres treated by micro-cutting or threading. However, the lateral intensity of the threaded fibres decreases drastically, so that almost no UV radiation can be measured from the middle of the activated area (Spot B and C). Fibres with micro cuts, on the other hand, can maintain a high intensity level between 90 and 100% of the initial intensity over the three measuring points (Spot A, B, C). This applies to each of the three fibres examined. The fluctuations that occur are due to the manual processing of the surfaces.

The commercial sidelight fibre also shows a very homogeneous radiation pattern. However, the intensity level is more than 60 percentage points lower compared to the fibres with micro cuts, so that the lateral range of the emitted radiation is significantly smaller. As expected, the untreated reference fibre shows no significant nor measurable sidelight intensity.

5 Adhesive Experiments

The previously gained knowledge from the characterisation of the fibres is subsequently validated in adhesive tests. With Photobond FB4175 and Loctite AA3525, two radical reacting UV adhesives are investigated. The product data have already been listed in Table 2. While the manufacturer recommends a wavelength of 400 nm for the FB4175, the AA3525 is designed for 365 nm. Thus, the validation tests can be used to investigate the influence of the wavelength on the fibre activation in parallel.

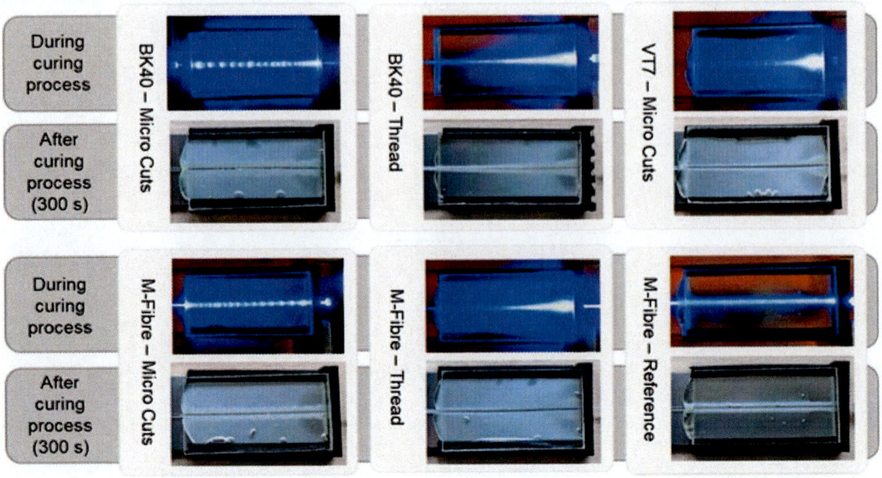

Fig. 8 Curing adhesive samples (DELO Photobond FB4175) by guiding UV-light (wavelength 400 nm) into POF for 300 s

The test specimen used for the validation is a visual sample whose joining partners consist of an aluminium sheet and a transparent PMMA. The transparent component is necessary to enable an optical evaluation of the light propagation and range of the adhesive curing. This allows conclusions to be drawn about the possible process limits for the later target application on non-transparent components. The length of the bonding surface is 50 mm, analogous to the length of the activated fibre area. The minimum adhesive layer thickness depends on the fibre diameter and is 1.0 mm in the tests performed. 3D printed polylactide (PLA) spacers also adjust the adhesive layer thickness and define the lateral edges of the bonding surface. Another PLA clip centers the fibre along the sample. Figure 8 shows the visual sample with the previously characterized fibres and the adhesive FB4175 during and after the curing process of 300 s at a wavelength of 400 nm.

During the curing process, the lateral light distribution along the test area is clearly visible. The distribution of the fibres processed with gradual micro cuts is much more homogeneous than the fibres with threads. In the latter case, most of the UV radiation is emitted in the initial area and decreases very strongly in the further course. The reason for this is the thread cut, which damages the fibre surface so much that the axial light conduction is completely blocked and all light is emitted laterally over a short distance. This also correlates with the results from lateral and axial intensity measurements. An exception is the VT7 fibre manufactured at ITA. Gaseous scattering centres in the fibre which are induced during the meltspinning process cause inhomogeneous lateral emission despite the gradually increasing arranged micro cuts. The untreated commercial sidelight fibre (M-Fibre) shows a very homogeneous lateral radiation characteristic. However, the intensity of the lateral light is significantly lower than that of the treated fibres, which was already noticed during the tests for fibre characterization.

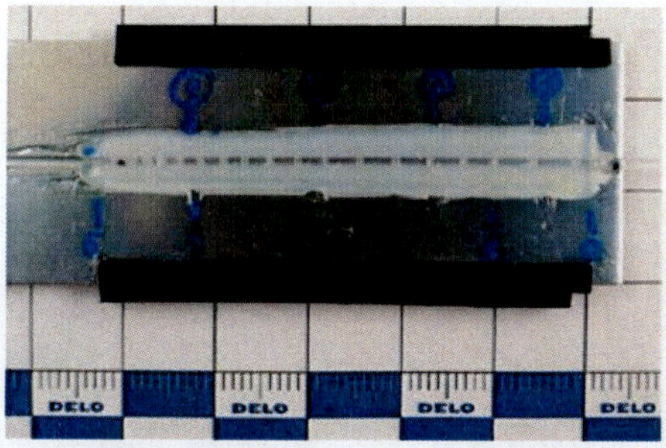

Measuring point	1	2	3	4
Width of the cured adhesive at measuring point [mm]	7.0	6.1	5.7	5.1
Distance of the measuring point for light injection [mm]	5	17	31	44
Progress of cure [%]	> 90	< 90	< 90	< 90

Fig. 9 Examination of the degree of curing by DELO Industrie Klebstoffe GmbH & Co. KGaA, Windach, Germany

After the curing process, the described observations on light distribution can be recognized in the degree of curing of FB4175. The adhesive discolours light and cloudy when cured. Darker, transparent adhesive areas have reacted less or not at all. The most uniform degree of cure was achieved by micro-cutting. Here, a distinction can be made between areas close to the fibre (<2.5 mm) and areas further away from the fibre (>2.5 mm). In the area <2.5 mm the curing is almost completely finished. In the area further away from the fibre, the maximum curing is not yet reached and the adhesive can still be removed from the substrate. This was also proven by infrared spectrometric measurements on the test specimen at the adhesive manufacturer DELO, see Fig. 9.

To investigate a second adhesive with a different wavelength, the described tests are additionally performed on the adhesive Loctite AA3525 and with UV light with a wavelength of 365 nm. The corresponding results are shown in Fig. 10.

The lateral radiation characteristic is the same at 365 and 400 nm. Again, the fibres with micro cuts lead to the most homogeneous light extraction. However, large differences between the two adhesives can be identified after the curing process. The investigated Loctite AA3525 adhesive cures in a clear and transparent state. When examining the images after an irradiation of 300 s, only fine edges can be identified in the form of the radiation pattern in the immediate proximity of the fibre. The lateral curing depth is therefore <1 mm. One reason for this is the wavelength used. Although UV light at 365 nm is at a higher energy level than at a wavelength of 400 nm, the attenuation in the light-conducting POF is also wavelength-dependent. In the range under consideration, a higher wavelength leads to lower attenuation and thus to less light and energy loss. The effective radiation energy is further reduced by the lateral emission from the fibre of the light. As a result, the energy level falls below the minimum required and the photo initiators in the UV adhesive cannot start the curing reaction. At the same time, the range of the radiation depends on its wavelength and is correspondingly smaller at 365 nm than at 400 nm. This contributes to the fact that at 365 nm the reactions in the UV adhesive only take place in close proximity to the fibre.

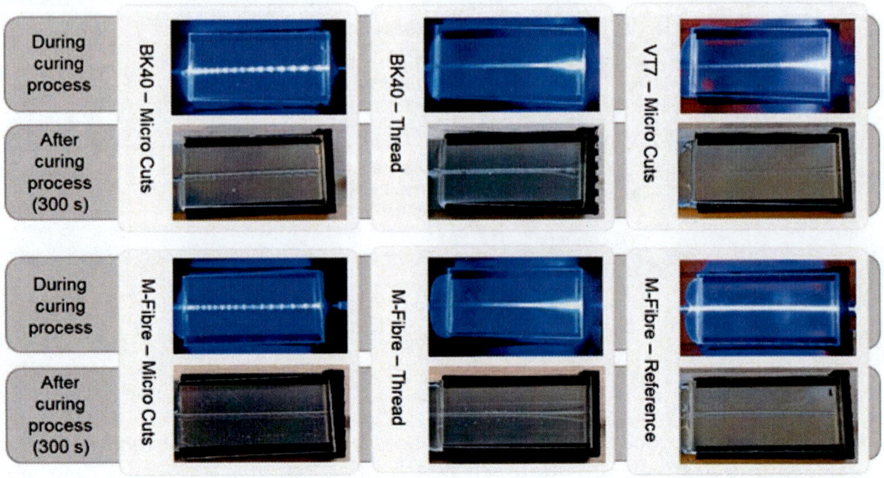

Fig. 10 Curing adhesive samples (Loctite AA3525) by guiding UV-light (wavelength 365 nm) into POF for 300 s

As a result, it becomes clear that a UV light source with 365 nm in the combination of the considered fibre material (PMMA) and the considered adhesive is not appropriate for the intended purpose. The curing of the adhesive Photobond FB4175 by side light of an activated POF at 400 nm wavelength delivers promising results. Curing of the adhesive could be reliably achieved for a range of up to 2.5 mm. To reduce the exposure time of 300 s, further investigations with alternative radiation sources (e.g. UV laser diodes) and an optimization of the light coupling between light source and fibre are planned. The thickness of the adhesive layer depends primarily on the diameter of the POF. POF with smaller diameters allow lower adhesive layer thicknesses. However, there is a quadratic relationship between fibre diameter and the light that can be coupled into the fibre via the fibre end face, so that halving the diameter leads to a reduction in possible light emission by a factor of four. This suggests as next steps to use several smaller fibres or POF woven fabrics as light guides between the bonding components.

6 Conclusions

In this paper, the concept of FibreKleb technology is explained and the procedure for identifying process limits is described. Polymer optical fibres (POF) of PMMA with a diameter of 1.0 mm are investigated. The fibre surfaces are activated with gradual micro cuts and threads so that a lateral light emission is achieved. Subsequently, the fibres are characterized by axial and lateral intensity measurements. The results are compared with those of a commercial sidelight fibre. Over the activated test length of 50 mm, the best result in terms of homogeneity and intensity of the side light is achieved by a treatment with micro cuts. These findings are validated on adhesive samples. Two UV adhesive systems are used, which are cured under irradiation of UV light with the wavelengths 365 and 400 nm. While lateral homogeneous adhesive layer depths of up to 2.5 mm are achieved at 400 nm, the use of UV light at 365 nm proves unsuitable for the selected tests.

Acknowledgements These presented results were developed within the framework of the research project FiberKleb, IGF-No. 20382 N. The project was funded by the Arbeitsgemeinschaft industrieller Forschungsvereinigungen "Otto von Guericke" e.V. (AiF) from funds of the Federal Ministry of Economic Affairs and Energy (BMWi) based on a decision of the German Bundestag and is supported by DVS - Deutscher Verband für Schweißen und verwandte Verfahren e.V.

References

1. Allied Market Research by Allied Analytics LLP (2020) UV adhesives market by resin type (acrylic, cyanoacrylate, epoxy, silicone, polyurethane, and others) and application (packaging, electronics, medical, transportation, industrial assembly, glass bonding, and others)—global opportunity analysis and industry forecast, Mar 2019–2026
2. Bunge C-A, Beckers M, Lustermann B (2017) Basic principles of optical fibres. In: Bunge CA, Gries T, Beckers M (eds) Polymer optical fibres, 1st edn. Woodhead Publishing, Sawston, England
3. Wellmann S, Messerig J, Nuyts N (2006) Lichthärtung auch im Schattenbereich. In: Adhäsion: Kleben & Dichten, Jahrgang 50, Heft 1–2. Springer Vieweg, Wiesbaden, Germany
4. Hartwig A (2012) Kationisch härtende Epoxidharzklebstoffe. Workshop Epoxidharz-Klebstoffchemie für Klebstoffanwender in Industrie und Handwerk. Frankfurt a. M., Germany
5. Amrhein F, Kehrer M, Schreck T, Bochtler U, Hellmann R, Kaloudis M (2011) UV-LED-Bestrahlung von Vergussmaterial: Schnelle Aushärtung auch im Schatten. In: Adhäsion: Kleben & Dichten, Jahrgang, Heft 5. Springer Vieweg, Wiesbaden, Germany
6. Pennig D (2019) Fotodynamische intramedulläre Knochenstabilisierung – neue Möglichkeiten in der Frakturbehandlung. Unfallchirurg 122:586. https://doi.org/10.1007/s00113-019-0683-z
7. ILLUMINOSS Medical, Inc. (2020) Fracture repair at the speed of light. https://www.illuminoss.com/. Retrieved 28.02.2020. Illuminoss Medical, Inc.
8. Kerkhoff J (2015) Blaues Wunder im OP. https://www.perspektiven-maga-zin.de/fileadmin/srh/perspektiven/Ausgabe_04_2015/03_gesundheit/02_Blaues_Wunder_im_OP/Blaues_Wunder_im_OP_pm_04_2015.pdf. Retrieved 07.11.2017. Perspektiven, Magazin für Bildung und Gesundheit, SRH Holding, Heidelberg
9. Alobaidani AD, Furniss D, Johnson MS (2009) Optimization and spectroscopic experiments of cationic epoxy formulations for the application of photocuring of deep resin mediums using side emitting optical fibres. In: Proceedings, international committee on composite materials, ICCM-17, Edinburgh, Scotland
10. Alobaidani A (2009) Development of a photocuring system for cationic epoxy formulations using side emit-ting optical fibres. Ph.D. thesis, University of Nottingham, United Kingdom
11. Shustack PJ, Logunov SL, Fewkes EJ, Dewa PG (2014) Photocuring in areas where you typically cannot get light. In: UV & EB technical conference proceedings. Radtech: The Association for UV&EB Technology, Chevy Chase, USA

12. Domaschk R, Päthe D (2005) Verfahren zum Verbinden von Bauteilen, insbesondere von Kfz-Karosserieteilen mittels eines Klebers, Innovationsgesellschaft für Fortgeschrittene Produktionssysteme in der Fahrzeugindustrie mbH. Application Number: DE19604081A; Publication Date: 07/28/2005
13. Beckers M, Schlüter T, Vad T, Gries T, Bunge C-A (2015) An overview on fabrication methods for polymer optical fibers. Polym Int 64:25–36

Some Industrial Examples of Accelerated Curing Using Curie Particles

Ahmed Elmahdy, Isabel Van de Weyenberg, Patrick Cosemans, Michael Adam, Morten Voß, Sascha Heinrichs, and Till Vallée

Abstract The aim of this paper is to assess the potential of accelerating the curing of epoxy adhesives by means of induction and ferromagnetic curie particles in the electronic and the automotive industries. Commercially available Araldite 2011 and Jowat 690 epoxy adhesives were considered, while Mn-Zn-Ferrite particles were considered for the modification of the adhesives. Induction experiments were performed at frequencies ranging from 188 to 262 kHz, and with particle weight content of 16.7 and 30%. The effects of the induction frequency and the particle content on the temperature of the adhesive were investigated. In addition, the effect of the particle content on the lap shear strength of bonded joints was also studied. Moreover, a comparison between the lap shear strength of inductively cured joints and reference oven cured joints was performed. Results showed that the temperature of the adhesive increased with increasing both the induction frequency and the particle content. However, the lap shear strength of joints with Araldite 2011 was reduced by 17% when increasing the particle content to 30%. No impact on the strength was observed in the case of the Jowat 690 at 16.7 and 30% particle contents. This suggests that increasing the induction frequency is more efficient in increasing the adhesive temperature without compromising the lap shear strength of the brittle adhesive. Furthermore, the inductively cured samples of the Araldite 2011 with 30% particles achieved a substantial 120% of the reference lap shear strength of oven cured samples in 8% of the curing time. For samples with Jowat 690 with 30% particles, the inductively cured samples achieved 43% of the reference lap shear strength of oven cured samples in 8% of the curing time. This shows a high potential

A. Elmahdy (✉) · I. Van de Weyenberg
Flanders Make, Core Lab Productions, Oude Diestersebaan 133, 3920 Lommel, Belgium
e-mail: ahmed.elmahdy@flandersmake.be

P. Cosemans
Sirris, Wetenschapspark 3, 3590 Diepenbeek, Belgium

M. Adam · M. Voß · S. Heinrichs · T. Vallée
Fraunhofer Institute for Advanced Manufacturing, Wienerstraße 12, 28359 Bremen, Germany

Lucas F. M. da Silva et al. (eds.), *Industrial Applications of Adhesives*,
Lecture Notes in Mechanical Engineering,
https://doi.org/10.1007/978-981-15-6767-4_2

of accelerating the curing time for industrial applications by induction and ferromagnetic curie particles, to reach either high strength as in case of Araldite 2011 or the handling strength as in case of Jowat 690.

Keywords Accelerated curing · Ferromagnetic particles · Epoxy adhesives · Lap shear strength · Induction heating

1 Introduction

Adhesive bonding using epoxy is currently considered a key technology in assembling multi-material parts and structures. Compared to classical bolted and riveted joints, adhesively bonded joints using structural epoxy adhesives offer a higher mechanical performance and a better weight reduction. Additionally, adhesive bonding offer the possibility to assemble dissimilar materials which are typically difficult to assemble mechanically, such as glass and composite materials. Despite the many advantages of epoxy adhesives, their relatively long curing time increase the overall manufacturing time and introduces a bottleneck in assembly lines. Some epoxy adhesives cure either at elevated temperatures or at room temperature. Depending on the type of the epoxy adhesive used, the whole assembly has to be either cured inside an oven, or stored for long times in an open inventory. This means that, besides the time requirement, unwanted thermal stresses in the structures can be introduced. In addition, extra costs of inventory could be added to the overall cost of the part. Consequently, it is important to reduce the curing times of structural adhesives.

Curing by induction is one of the techniques which is recently used to reduce the curing time of structural adhesives. Induction heating itself is a well-known technique used to heat and melt metallic materials. In this technique, metallic materials (either electrically conductive or magnetic) are exposed to a high intensity alternating magnetic field. The heat is generated either by the joule effect (resulting from the flow of the alternating eddy currents in the material), or by hysteresis losses (resulting from the internal friction of the molecules due to the change of the magnetic polarity). The induction technique is used to cure adhesives in component assembly of electric motors [1], and in several automotive applications [2]. In such applications, the metallic adherent material (either ferroelectric or ferromagnetic) heats up by means of induction, and the heat is transferred to the adhesive via thermal conduction. However, if the adherent material is nonconductive or non-ferromagnetic, the adhesive has to be modified with some susceptor particles in order to generate heat within the adhesive.

The modification of polymers and adhesives using susceptor particles for induction heating was studied by several researchers. Bayerl et al. [3] studied the influence of induction frequencies, particle type, and particle content on the heating and the mechanical behavior of polyethylene and polyamide thermoplastics. Results showed that induction frequency of 2500 kHz was not sufficient to heat the thermoplastic/particle mix. Induction frequency of 450 kHz showed good heating effects, where the melting temperature of both thermoplastics could be reached in 3 min. In addition, the ferromagnetic particles (iron and magnetite) showed a better heating effect compared to carbon black particles which is only electrically conductive. Impact strength was also significantly reduced with the increase of particle content. Vallée et al. [4] investigated the effect of adding magnetite particles to Mn-Zn-ferrite particles in 1-component epoxy for induction heating at 373 kHz. Results revealed that the addition of 4% magnetite helped in increasing the adhesive temperature. However, subsequent increase in magnetite content up to 12% did not have additional influence on the adhesive temperature. Severijns et al. [5] studied the influence of iron particle content on the lap shear strength of bonded joints. Results showed that the lap shear strength was reduced by 15% with the addition of 0.5% volume content of particles compared to unfilled adhesive. In addition, the inductively cured joints showed 6% higher lap shear strength compared to oven cured joints.

This study aims to assess the potential of accelerating the curing of commercial epoxy adhesives using induction heating and curie particles, especially in electronics and automotive industries. Two industrial applications are considered in this study. The first application is potting of electronic sensors in a glass fiber reinforced polypropylene sulfide composite (PPS) housing. These sensors are typically used in monitoring the air quality in heat, ventilation, and air conditioning (HVAC) systems. Two-component epoxy adhesives are generally used for this process as they protect the sensors against humidity, high temperatures, and corrosion [6]. The potted assembly is typically cured in the oven at relatively low temperatures for longer hours (up to 80 °C for 12 h) so as not to damage the electronics and the housing materials. The second application is the bonding of a multifunctional bracket onto a glass windshield panel of passenger vehicles. The multifunctional bracket acts as an anchor point for the rearview mirror and additional driver assistance modules such as cameras, light sensors, rain sensors, etc. The bracket is typically made of glass fiber reinforced PBT/PET composite, while the surface of the laminated glass is treated with thin layer of enamel. Flexible epoxy grade adhesives and polyurethanes are generally used for this application to reduce stress concentrations and to damp vibrations [7]. The bonded assembly is generally cured at room temperature for 48 h. In case of curing at elevated temperatures, the maximum allowable curing temperature is also 80 °C in order not to damage the polyvinyl butyral (PVB) foil in the laminated windshield glass.

In this paper, the effect of the induction frequency and the weight percentage of the curie particles on the temperature in the adhesive is first investigated. Secondly, the effect of the weight percentage of curie particles on the mechanical performance of the adhesively bonded joints is studied using standard single lap shear samples. Finally, the potential of accelerating curing using induction is assessed and verified on single lap shear samples with regard to the abovementioned applications. The mechanical performance of the inductively cured joints is benchmarked against reference oven cured joints.

2 Materials and methods

2.1 Selection of adhesives and curie particles

Two epoxy based adhesives were considered in this study. For the sensor potting application, the Araldite 2011 two component epoxy (manufactured by Huntsman) was selected. This adhesive provides full electric insulation of the electronic sensors, high temperature resistance and good bonding with the adherent material. Additionally, no volatile contents are produced during curing. For the multifunctional bracket/windshield assembly, the Jowat 690 two component SE polymer was selected. This adhesive is a hybrid of epoxy resin and silane terminated polymer, which offers good flexibility, good bonding with glass, and solvent free curing. According to the Araldite 2011 datasheet, the recommended minimum curing time at 70 °C is 50 min. For the Jowat 690, the minimum curing time recommended by the datasheet is 24 h to reach 50% of the final strength. Increasing the adhesive temperature could reduce the required curing time, however, no high temperature curing cycles were specified in the data sheet. Full curing is achieved after 7 days at room temperature. Given the selected curing temperature for both applications as mentioned in the introduction section, Dynamic Scanning Calorimetry (DSC) measurements were performed to determine the minimum curing time of both adhesives at 80 °C. Figure 1 shows the evolution of the conversion fraction (α) at isothermal curing conditions at 80 °C for 150 min for both adhesives, considering a heating rate of 50 °C/min. As seen in Fig. 1, the Araldite 2011 achieved a conversion of approx. 99% after 60 min, while the Jowat 690 achieved only 70% conversion after 150 min. By extrapolating the conversion curve of the Jowat 690, a conversion of 90% can be achieved within 5–6 h at 80 °C.

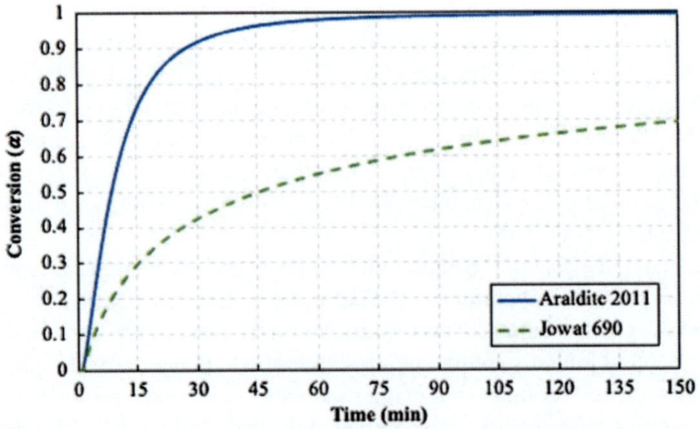

Fig. 1 Evolution of the conversion fraction (α) with time for Araldite 2011 and Jowat 690 epoxy based adhesives at 80 °C isothermal curing conditions

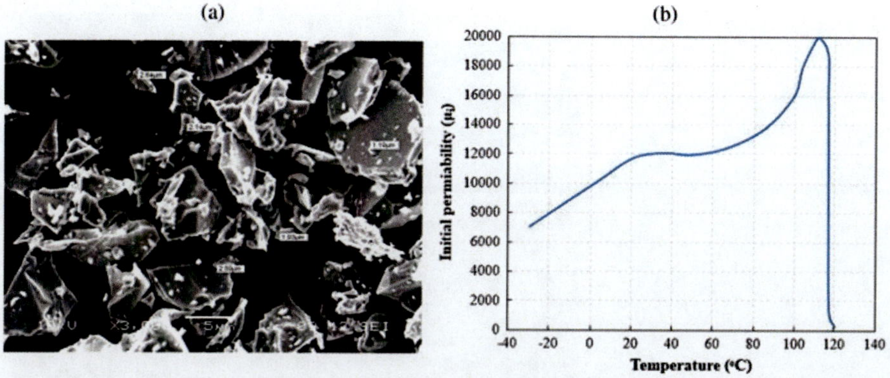

Fig. 2 **a** SEM image of the R12K particles (magnification 3000×), **b** evolution of the initial permeability or the R12K particles with temperature [8]

The selected curie particles were Mn-Zn-ferrite R12K particles (supplied by Hengdean group DMEGC Magnetics, China). The particles were supplied in one batch in a plate like morphology, with a mix of sub-micrometer and micrometer sizes having several sharp edges. An SEM analysis was performed to determine the average size and distribution of the R12K curie particles, as shown in Fig. 2a. Approximately 50% of the particles had an average size of 0.5 μm, while the other 50% of the particles had an average size of 4.5 μm. According to the manufacturer's datasheet, the density and resistivity of the particles are 4.9 g/cm^3 and 0.15 Ω m respectively. The curie temperature of the particles is 110 °C, which is below the softening temperature of the adherent materials of both applications. This curie temperature corresponds to the temperature at which the initial magnetic permeability of the particles drops to zero. Figure 2b shows the evolution of the initial permeability with temperature.

2.2 *Induction setup and temperature measurement*

Induction experiments were performed on the TruHeat HF 1005-5010 induction machine (manufactured by TRUMPF Hüttinger GmbH) available at the Joining and Materials Lab at Flanders Make. It consists of 3 main units: a water cooling unit, a generator unit, and an inductance-capacitance circuit unit with the induction coil. The power generator is capable of supplying a maximum induction power of 11.2 kW and a maximum current of 35 A. Figure 3 shows the induction setup used. The main specifications of the induction machine are summarized in Table 1.

Three flat spiral coils (pancake coils) made up of 4 turns were used as shown in Fig. 4. The coils were manufactured from hollow copper tubes (with outer diameters 8, 6, and 4 mm respectively), which allowed the flow of the cooling water inside the inductors. The frequency ranges achieved by the coils were 180–228 kHz for the large coil, 196–252 kHz for the medium coil, and 243–325 kHz for the small coil.

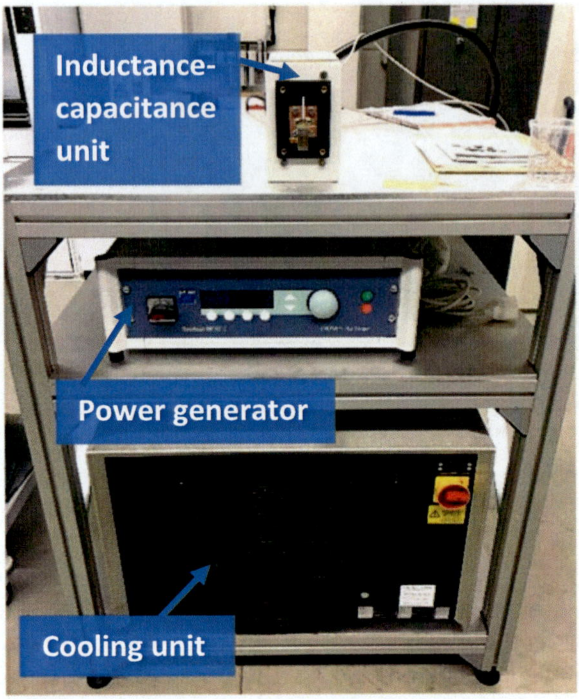

Fig. 3 Induction setup used

Table 1 Specifications of the induction machine

Parameter	Value
Transformer ratio	16:1
Capacitors (nF)	660
Frequency range (kHz)	50–1000
Current range (A)	2.8–35
Maximum voltage (V)	1200
Maximum power (kW)	11.2

Temperature inside the adhesive and in the bond line was measured using a partial immersion liquid-in-glass thermometer, having a measurement range of −10 to 250 °C. The use of liquid-in-glass thermometer ensured the accurate measurement of the temperature in the bond line without any interferences due to the magnetic field.

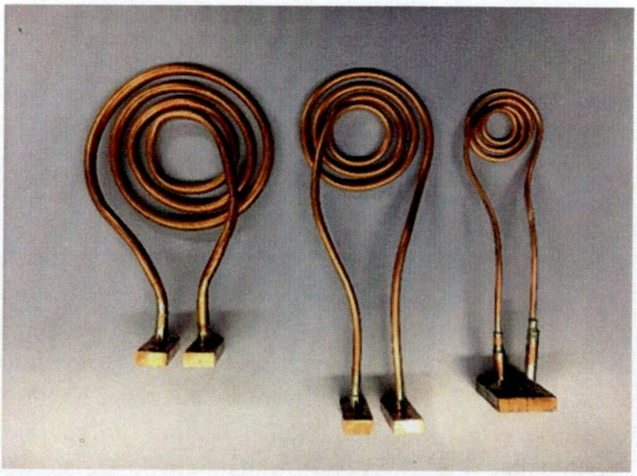

Fig. 4 Flat spiral (pancake) coils used

2.3 Single lap shear test setup

Single lap shear tests were performed using a Shimadzu AGS-50NX universal testing machine shown in Fig. 5. The load was measured using a 20 kN load cell. Special attention was paid to ensure the alignment of the lap shear samples with the grips in order to eliminate any bending moments. Samples were tested at cross head speeds of 2 mm/min for the joints with Araldite 2011 adhesives, and 5 mm/min for the joints

Fig. 5 Single lap shear testing setup

with Jowat 690 adhesive, according to the recommendations of the ASTM D3163 standard [9].

2.4 *Specimen materials and geometries*

Adhesive/particle mixtures were prepared by manually mixing the R12K particles with each adhesive separately with particle weight percentages of 16.7% and 30% respectively. Several single lap shear samples were prepared and cured using the adhesive/particle mixtures and neat adhesives for the different investigations, as summarized in Table 2. The adherent materials were Ryton® polypropylene sulfide (PPS) reinforced with 40% glass fibers, Ultradur® polybutylene terephthalate/polyethylene terephthalate (PBT/PET) reinforced with 30% glass fibers, and glass coated with thin enamel layer on the adhesion side. The length and the width of the adherent materials were 100 mm and 25 mm respectively. The thickness of the adherent materials were 2 mm for the PPS and PBT/PET, and 5 mm for the glass. The surfaces of the

Table 2 Summary of the prepared single lap shear samples

Set number	Adhesive mixture	Adherent materials	Curing conditions	Required study
1	Neat Araldite 2011	PPS/PPS	Oven cured for 3 h at 80 °C	Effect of particle content on the lap shear strength
2	Araldite 2011 + 16.7% R12K	PPS/PPS	Oven cured for 3 h at 80 °C	
3	Araldite 2011 + 30% R12K	PPS/PPS	Oven cured for 3 h at 80 °C	
4	Neat Jowat 690	PBT/Glass	RT cured for 48 h	
5	Jowat 690 + 16.7% R12K	PBT/Glass	RT cured for 48 h	
6	Jowat 690 + 16.7% R12K	PBT/Glass	RT cured for 48 h	
7	Araldite 2011 + 30% R12K	PPS/PPS	Oven cured for 3 h at 80 °C	Thick adherent samples for temperature monitoring in bond line during induction
8	Jowat 690 + 30% R12K	PBT/PBT	RT cured for 48 h	
9	Araldite 2011 + 30% R12K	PPS/PPS	Induction curing	Application assessment
10	Jowat 690 + 30% R12K	PBT/Glass	Induction curing	
11	Jowat 690 + 30% R12K	PBT/Glass	Oven cured for 3 h at 80 °C	

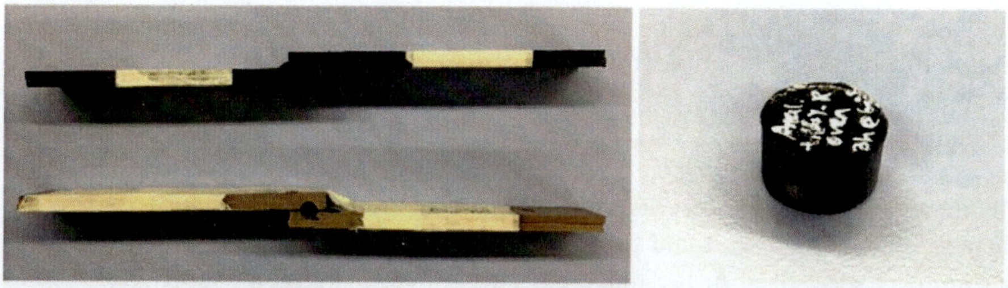

(a) (b)

Fig. 6 Samples for temperature measurement: **a** thick adherent lap shear samples with PBT/PBT sample at the top and PPS/PPS sample at the bottom, **b** cylindrical samples

thermoplastic composite samples to be bonded were treated with atmospheric pressure plasma prior to the application of the adhesive in order to improve the adhesion properties on the surface. All samples were prepared in special molds to maintain a bond line thickness of 0.1 mm, and an overlap distance of 12.5 mm, according to the ASTM D3163 standard [9]. The oven cured samples were left to cool in the oven for 16 h after curing. Next to the lap shear samples, small cylindrical samples of diameter 18 mm and height of 10 mm were prepared to study the temperature in the adhesive at different induction frequencies and different particle weight content. The samples were prepared using the Araldite 2011/R12K mixtures and cured in oven for 3 h at 60 °C. A blind hole of diameter 5.5 mm and depth 8 mm was drilled in each sample to accommodate the liquid-in-glass thermometer. Additionally, thick adherent single lap shear specimens (PPS/PPS and PBT/PBT) of adherent thickness 4 mm were also prepared using both adhesive mixtures in order to monitor the temperature in the bond line during induction, as indicated in Table 2. Similar to the disk samples, a blind hole of depth 8 mm was also drilled in the center of the overlap distance to accommodate the thermometer. Figure 6 shows the cylindrical and the thick adherent samples used.

2.5 Experimental Conditions

In order to study the effect of the frequency and the particle weight content on the adhesive temperature, the small cylindrical samples were inductively heated for 15 min with the 3 pancake coils. For all the coils, the induction power was kept constant at 0.5 kW. The induction frequencies used were 188 kHz for the large coil, 207 kHz for the medium coil, and 262 kHz for the small coil. The coupling distance between the coils and the samples were 5 mm. The temperature was manually recorded every 0.5 min using the liquid-in-glass thermometer.

The effect of particle content on the lap shear strength was studied using oven cured samples of the Araldite 2011 adhesives, and room temperature cured samples for the Jowat 690, as detailed in Table 2. Three particle weight percentages of 0%, 16.7%, and 30% were tested. At least 3 samples were tested for each testing condition. The tests were conducted at ambient temperature of approx. 22 °C and relative humidity of approx. 34%

For the induction curing tests, the lap shear samples where inductively cured using the small pancake coil for 15 min, at 5 kW induction power and 245 kHz induction frequency. All samples were fixed in a special fixture during induction, where the distance between the coil and the centerline of the bond line was 7 mm. Figure 7 shows the fixture used in the induction curing tests. The temperature in the bond line during induction was monitored using the thick adherent samples. The fixture of these sample were adjusted such that the distance between the coil and the bond line was also 7 mm. The selected induction parameters achieved a curing temperatures of approx. 80 °C after 15 min. Figure 8 shows the induction power cycle and the temperature developed in the bond line during the induction curing tests. The lap shear strength of the inductively cured samples for both adhesive mixtures were compared against the oven cured samples of the same adhesive mixture. The oven curing conditions were 3 h at 80 °C, which is the same curing temperature achieved by the induction curing. All induction samples were left to cool for 1 h before lap shear testing. The tests were also conducted at ambient temperature of approx. 22 °C and relative humidity of approx. 34%.

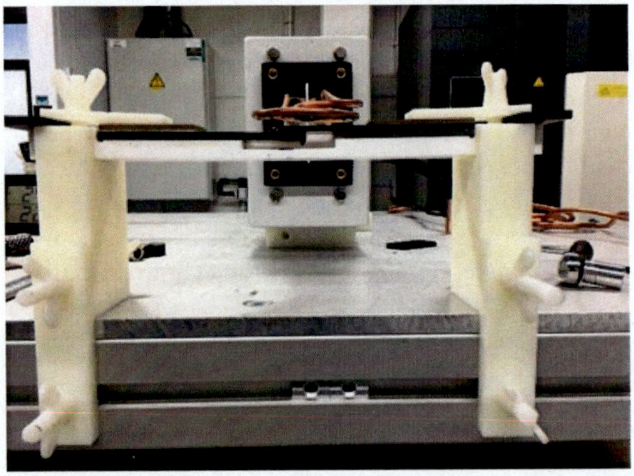

Fig. 7 Sample fixture used during induction

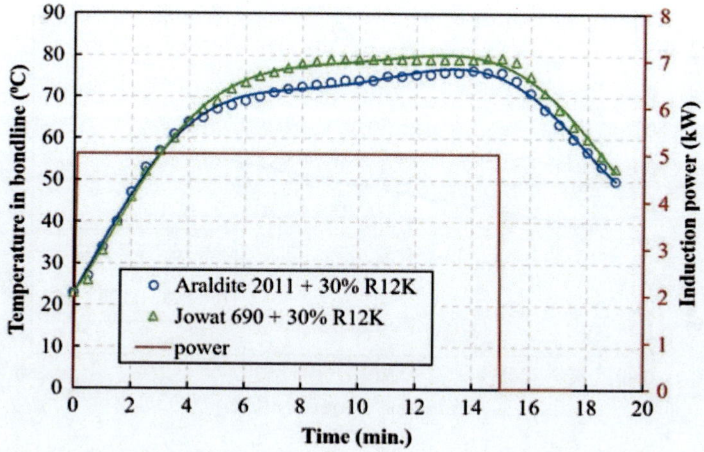

Fig. 8 Induction power cycle and temperature in the bond line during induction

3 Results and discussion

3.1 Effect of frequency and particle content on the adhesive temperature during induction

Figure 9 shows the evolution of the temperature in the araldite 2011 adhesive at different induction frequencies and particle contents for 15 min. It can be seen that the temperature increased gradually for all induction frequencies and particle contents, and reached a stable level after approx. 13 min. The temperature generally increased with the increase of both the induction frequency and the particle content in the adhesive.

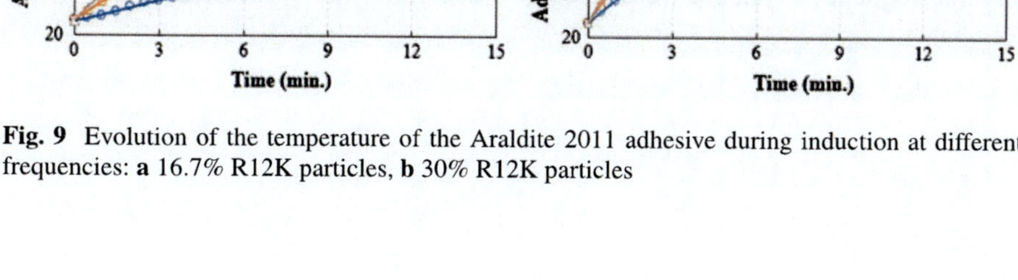

Fig. 9 Evolution of the temperature of the Araldite 2011 adhesive during induction at different frequencies: **a** 16.7% R12K particles, **b** 30% R12K particles

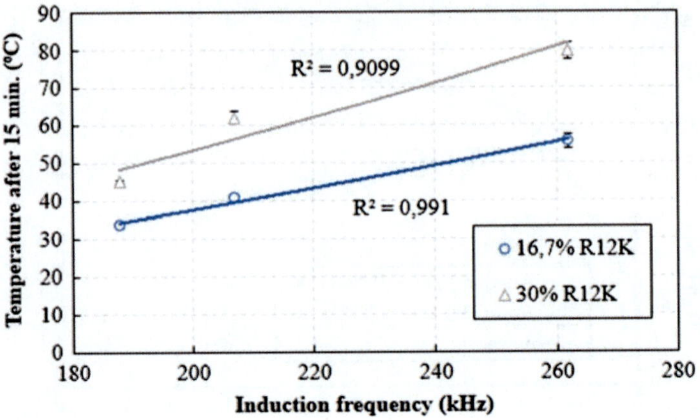

Fig. 10 Temperature of the araldite adhesive with different R12K particle content at different induction frequencies after 15 min induction

Figure 10 shows the temperature of the adhesive mixtures as a function of the induction frequency after 15 min induction. For the adhesive with 16.7% particle content, the temperature increased from 34 °C at 188 kHz to 56 °C at 262 kHz, which corresponds to an increase of approx. 65%. For the adhesive with 30% particle content, the temperature increased from 45 °C at 188 kHz to 79 °C at 262 kHz, which corresponds to an increase of approx. 75%. On the one hand, increasing the particle content by approx. 80% at the same induction frequency only increased the temperature in the adhesive by percentages of 34–52%. On the other hand, increasing the induction frequency by just 39% at the same particle content can increase the temperature by percentages of 65–75%. This suggests that increasing the induction frequency is more efficient in raising the adhesive temperature compared to increasing the particle content. This is attributed to the pronounced skin effect during induction heating. By using a least square power fit, the increase in adhesive temperature with the increase in induction frequency can be represented by the following relation:

$$T_A = Kf^m \tag{1}$$

where T_A is the adhesive temperature, K is the temperature coefficient in °C, and m is the frequency exponent. For the adhesive with 16.7% particle content, the temperature coefficient and frequency exponent were 0.0147 °C and 1.4808 °C respectively. For the adhesive with 30% particle content, the temperature coefficient and the frequency exponent were 0.0118 °C and 1.5886 °C respectively.

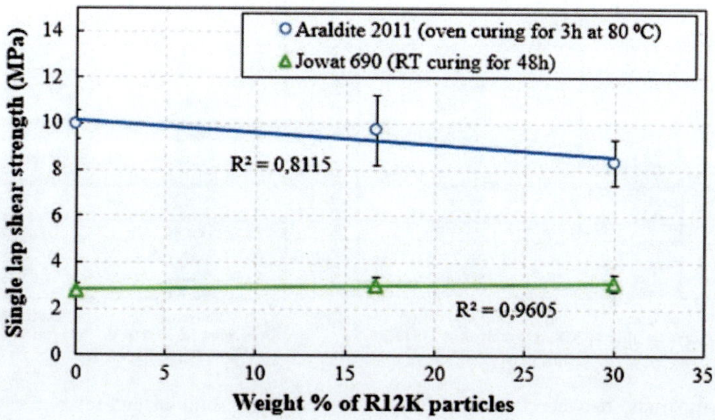

Fig. 11 Effect of particle content on the lap shear strength using Araldite 2011 and Jowat 690 adhesives

3.2 *Effect of particle content on the lap shear strength of both adhesives*

Figure 11 shows the effect of particle content on the lap shear strength using both Araldite 2011 and Jowat 690 adhesives. It can be seen that for the Araldite 2011, the increase in the particle content decreased the lap shear strength. At 30% particle content, the lap shear strength was reduced to approx. 8.3 MPa compared to 10 MPa for the neat adhesive. This corresponds to a percentage reduction of approx. 17%. For the Jowat 690, the lap shear strength was almost constant at all particle contents. The reduction in the lap shear strength of the Araldite 2011 is attributed to the effect of stress concentrations caused by the particles in the brittle adhesive. Whereas in the case of the flexible Jowat 690, the stress concentration effects caused by the particles are much less pronounced.

3.3 *Assessment of the induction curing technique for the selected applications*

Figure 12a shows a comparison between the lap shear strength of the inductively cured PPS/PPS lap shear sample for 15 min and a reference oven cured sample (3 h at 80 °C) using the araldite 2011/30% R12K adhesive mixture. It can be seen that the inductively cured samples achieved a lap shear strength of approx. 12 MPa compared to approx. 8 MPa for the oven cured sample in just 15 min. This corresponds to a percentage of 150% of the reference lap shear strength in only 8% of the curing time. Similar results were reported by Severijns et al. [5], however, further research is still required to understand why inductively cured lap shear samples with the araldite

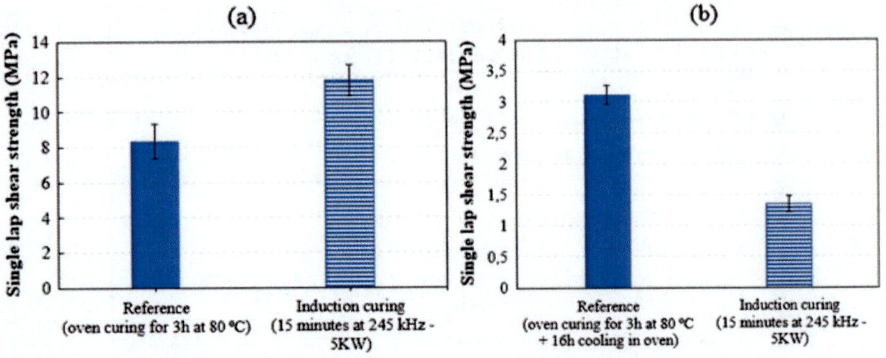

Fig. 12 Comparison between the inductively cured lap shear samples and reference oven cured samples: **a** potting application with Araldite 2011, **b** automotive windshield application with Jowat 690

2011 adhesive increased their lap shear strength compared to oven curing at the same curing temperature.

Figure 12b shows a comparison between the lap shear strength of the inductively cured PBT/Glass lap shear sample for 15 min and a reference oven cured sample (3 h at 80 °C) using the Jowat 690/30% R12K adhesive mixture. Contrary to the Araldite 2011 case, the inductively cured samples achieved a lap shear strength of 1.34 MPa compared to 3.1 MPa for the oven cured samples. This corresponds to a percentage of only 43.2% of the reference lap shear strength in 8% of the time.

4 Conclusion

This study was performed in order to assess the potential of accelerating the curing of epoxy adhesives by induction and curie particles in two industrial applications. The first application was potting of electronic HVAC sensors in a thermoplastic composite housing. The second application was the assembly of a multifunctional thermoplastic bracket in glass windshields of passenger vehicles. Two adhesives were considered in this study: Araldite 2011 2-component epoxy for the electronic potting, and Jowat 690 flexible epoxy/SE polymer adhesive for the multifunctional bracket. The effect of the induction frequency and the particle content on the temperature in the Araldite 2011 adhesive was investigated using a liquid-in-glass thermometer and an industrial grade induction machine. The range of the induction frequencies studied was 188 kHz up to 262 kHz, and the weight percentages of the particles were 16.7% and 30%. In addition, the effect of the particle content on the lap shear strength of both adhesives was studied using a universal tensile testing machine. Moreover, and with regard to the selected industrial applications, the lap shear strength of inductively cured samples of both adhesives was tested and compared against oven cured samples.

Considering the tested adhesives and samples, the induction conditions and the curie particles selected, the following can be concluded.

1. Increasing both the particle content and the induction frequency increases the temperature in the araldite 2011 adhesive.
2. Increasing the induction frequency by 39% at the same particle content can increase the temperature in the araldite 2011 adhesive by percentages of 65–75%, as compared to percentages of 34–52% if the particle content increased by 80%. This indicates that increasing the induction frequency is more efficient in raising the adhesive temperature compared to increasing the particle content.
3. The addition of 30% the particle content in the brittle Araldite 2011 adhesive reduces the joint strength by 17%.
4. The increase in the particle content does not have a significant effect on the joint strength of Jowat 2011 adhesive.
5. The inductively cured Araldite 2011/30% R12K joints achieved 150% of the reference oven cured strength in 8% of the oven curing time.
6. The inductively cured Jowat 690/30% R12K joints achieved 43.2% of the reference oven cured strength in 8% of the oven curing time.

Despite the lower strength achieved by induction for the Jowat 690/30% R12K, and given the fact that the Jowat 690 is a slow curing adhesive, it can still be considered that the handling strength was reached in 15 min. This handling strength is typically reached after 24 h according to the data sheet. The induction technology, therefore, shows a high potential of accelerating the curing time for industrial applications by modifying the adhesive ferromagnetic curie particles.

Acknowledgments This research was conducted as a part of the FACIDO Project: Fast and Stable Adhesive Curing with Debonding Option. The project is partially funded by Flanders Make (the strategic research center for the manufacturing industry) and the Flemish Agency for Innovation and Entrepreneurship VLAIO through the CORNET framework (Grant number: HBC.2016.0668). The authors acknowledge the support of NiniX Technologies Belgium, and AGC-Glass Europe, Belgium.

References

1. Arnosky J (2001) Induction heating in the manufacture of electric motors. Proceedings: Electrical Insulation Conference and Electrical Manufacturing and Coil Winding Conference. (Cat. No.01CH37264) Cincinnati, OH, USA, 57–62. https://doi.org/10.1109/EEIC.2001.965589
2. Rudnev V, Loveless D, Cook RL (2018) Handbook of induction heating, 2nd edn. CRC Press
3. Bayerl T, Schledjewski R, Mitschang P (2012) Induction heating of thermoplastic materials by particulate heating promoters. Polym Polym Compos 20:333–342. https://doi.org/10.1177/096739111202000401
4. Vallée T, Adam M (2016) Inductively cured glued-in rods in timber using Curie particles. Int J Adhes Adhes 70:37–45. https://doi.org/10.1016/j.ijadhadh.2016.05.005
5. Severijns C, de Freitas ST, Poulis JA (2017) Susceptor-assisted induction curing behaviour of a two component epoxy paste adhesive for aerospace applications. Int J Adhes Adhes 75:155–164. https://doi.org/10.1016/j.ijadhadh.2017.03.005

6. Bardoliwalla DF (1997) Fast curing, low exotherm epoxy potting and encapsulating systems, Proceedings: Electrical Insulation Conference and Electrical Manufacturing and Coil Winding Conference, Rosemont, IL, USA, 245–247. https://doi.org/10.1109/EEIC.1997.651053
7. Dunn DJ (2004) Engineering and structural adhesives 15:1 iSmithers Rapra Publishing ISSN:0889-3144
8. R12K material characteristics datasheet, Hengdian Group DMEGC magnetic limited, China
9. ASTM D3163-01(2014) Standard test method for determining strength of adhesively bonded rigid plastic lap-shear joints in shear by tension loading, ASTM International, West Conshohocken, PA. https://www.astm.org

Novel Torsion Machine to Test Adhesive Joints

M. A. Dantas, R. J. C. Carbas, A. M. Lopes, C. M. da Silva, E. A. S. Marques, and L. F. M. da Silva

Abstract Adhesive joints are increasingly being used in structural applications due to their improved mechanical performance when compared to other classical mechanical joining methods. In order to predict the mechanical performance of adhesive joints, it is first necessary to determine the mechanical properties of the materials being used (e.g. adhesives and adherends). These properties can be characterized under three basic loading modes: tensile, compression and shear. Tensile and shear tests are commonly used to determine the Young's modulus, tensile strength, and the strain to failure of materials. However, shear tests under torsion loading have been shown to provide significantly higher accuracy in the measurement of the strain to failure since no stress concentrations occur in the specimen. To explore the advantages associated with shear testing, a novel torsional testing machine was developed. It allows for accurate determination of the mechanical properties of structural adhesives by ensuring correct alignment of the specimens and avoiding any spurious bending moments, compression or tension loads during the tests. Different structural adhesives were selected and characterised under shear loads using this novel machine and the results obtained were compared with those obtained with thick adherend shear tests (TAST) results. This work showed that by performing torsion tests in the newly developed testing machine, the adhesive is properly characterised, especially for the determination of the strain to failure.

Keywords Structural adhesives · Shear properties · Torsion machine · Butt joint

M. A. Dantas · R. J. C. Carbas (✉) · E. A. S. Marques · L. F. M. da Silva
Instituto de Ciência e Inovação em Engenharia Mecânica e Engenharia Industrial (INEGI), Porto, Portugal
e-mail: rcarbas@fe.up.pt

R. J. C. Carbas · A. M. Lopes · C. M. da Silva · E. A. S. Marques · L. F. M. da Silva
Departamento de Engenharia Mecânica, Faculdade de Engenharia (FEUP), Universidade do Porto, Porto, Portugal

Lucas F. M. da Silva et al. (eds.), *Industrial Applications of Adhesives*,
Lecture Notes in Mechanical Engineering,
https://doi.org/10.1007/978-981-15-6767-4_3

1 Introduction

Adhesive joints are increasingly being used to replace mechanical joints due to their ability to join different types of materials and at the same time, maintain a more uniform stress distribution along the bonded connection. Adhesives generally allow for higher flexibility in structure design and represent a more cost-effective solution for most applications. Therefore, adhesive technology is now extensively being used in a wide range of industries, often optimised in terms of mechanical properties, sealing capability and temperature resistance for each specific case. This technology now finds use from civil construction to the aerospace industry [1].

In order to predict the behaviour of the adhesives and their failure under different types of loads, the mechanical properties of the adhesive materials must be first experimentally determined. Thus, a variety of different test procedures has been proposed to determine strength to failure and strain to failure of the adhesive's material. There are two main types of tests that can be performed for characterizing an adhesive, as the mechanical properties of the adhesive can either be determined using bulk specimens or in situ joints. Dolev and Ishai [2] found a good correlation between the in situ and the bulk properties of shear yield strength and elastic modulus derived from torsion tests. Lilleheden [3] has shown that the basic elastic properties can be evaluated trough tensile tests on bulk specimens. However, da Silva and Adams [4] have shown that although the elastic properties can be determined using this simple test, the complete shear curve cannot be deduced from the tensile curve, given the higher ductility of the adhesive in shear than in uniaxial tension. Dean et al. [5], Bryant and Dukes [6] and Chai [7] have all obtained good results, thus higher shear strength and strain to failure, when testing adhesives under torsion, either on butt joints or napkin-ring specimens. These results highlight the fact that a torsion machine is extremely useful for the mechanical characterization of adhesives. However, most of the equipment used in these works is based on commercial testing equipment which is not specifically designed to test adhesive joints and it is thus expectable that a custom designed equipment will be able to provide improved measurement capabilities for adhesively bonded specimens.

2 Techniques for the Mechanical Characterization of Adhesives Under Tension and Compression

Due to their simplicity and the ubiquity of universal testing machines, tensile and compression tests are still the most commonly used test procedures to mechanically characterize adhesives. In addition, the specimens and tests methods employed are generally comparable to those used for polymeric materials [8]. These tests are typically performed with bulk specimens, which are relatively easy to manufacture and allow to obtain properties which are fully independent from the substrate material.

2.1 Bulk Tensile Specimens

Tensile bulk specimens are one of the most common types of tests to determine adhesive strength and stiffness properties. As bulk specimens are composed only of a single material, the adhesive material, the properties obtained with a tensile test will be fully independent of the adherent materials which might not be the case when considering in situ tests. Bulk specimens are generally obtained by casting the adhesive in a mould or creating homogeneous plates by applying hydrostatic pressure to an uncured adhesive, as described by Banea et al. [9]. In the case of adhesives with low viscosity, preparing the specimens by injection is the most popular technique as it can be done with a mould and a syringe using a one-part adhesive or a previously mixed two-part adhesive. However, when working high viscosity, thixotropic adhesives it is often not possible to inject the material into a mould, with the use of hydrostatic pressure being the only suitable method to obtain an adhesive plate, which is then machined to the intended dimensions. The geometry generally used for the specimens follows the EN ISO 527-2 standard, with a *dog-bone* shape using longer specimens for rigid adhesives (e.g., epoxies) and short specimens for more flexible adhesives (e.g., polyurethanes) [8]. Plate specimens are frequently used in the industry as this type of tests are easier to produce given their simpler geometry, when compared with the previously mentioned specimens. These plates are produced according to the standard NF T76-142. da Silva et al. [10] obtained satisfactory results by using plate specimens for both film and paste adhesives, nevertheless, void formation still occurs during the manufacturing of these specimens.

Uniaxial tensile stress tests allow to determine the Young's modulus, the tensile strength and the failure strain from a single tensile stress–strain curve, Fig. 1. During these tests, environmental conditions such as temperature and humidity must be controlled given their significant influence on final obtained stress–strain curve, as shown by Banea and da Silva [11] for a different set of adhesives.

EN ISO 527-2 standard defines a method to determine the Young's modulus (E) by considering points at 0.05 and 0.25% strain, as shown in Fig. 1. Nevertheless, in the case of polymers, 0.25% may be already on the plastic strain zone because non-linear behaviour appears even for small deformations, thus compromising the

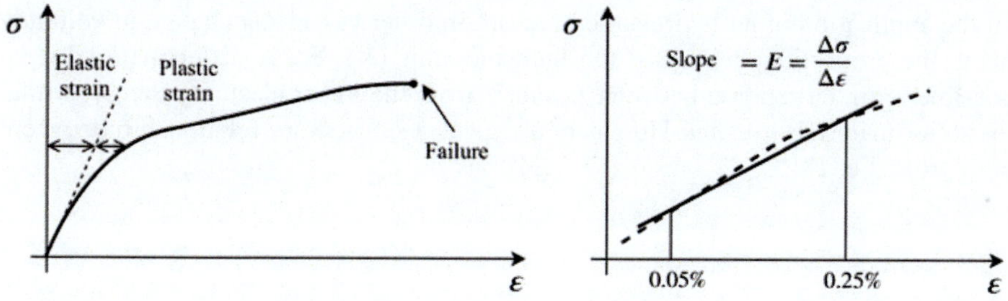

Fig. 1 Typical tensile stress–strain curve of an adhesive and methodology to determine Young's modulus. Adapted from: [8]

obtained results. Another property obtained in this test is the yield stress of the material (σ_y) and the peak of the stress–strain curve (σ_r). Because there is not a clear transition between the linear elastic region and the plastic region, various methods have been proposed to determine the yield point. ASTM D638-14 defines the yield point as the first point on the stress–strain curve for which an increase in strain occurs without an increase in stress. Furthermore, it is said that only materials whose stress–strain curves exhibit a point of zero slope may be considered as having a yield point.

To conclude, tensile stress specimens are convenient to determine the referred mechanical properties of adhesives. On the other hand, it is known that this type of tests presents very large dispersion in the obtained results given the existence of defects (mainly voids) in the specimens. The existence of a void in the material will induce a significant stress concentration factor on that zone and cracks will appear as a result, leading to premature failure of the specimen, well below the theoretical capabilities of the material. Therefore, strain to failure is a property which is highly dependent on the existence of defects in the specimens [1, 12].

2.2 *Bulk Compression*

Compression testing of the adhesive (uniaxially or hydrostatically) is not as common as tensile loading. However, for some cases compression stresses can appear due to the geometry of the joints and the applied loads. Testing bulk specimens under compression can be divided into two different cases: uniaxial compression and hydrostatic compression. Adhesive strength properties in compression and in tension are not the same since the compressive strength takes in account the hydrostatic component. da Silva et al. [1] refer that the ratio between the yield stress in compression and the yield stress in tension varies typically between 1.2 and 1.4.

Uniaxial compression tests are particularly suitable for metallic and polymeric materials and allow to obtain a stress–strain curve of the material. For soft materials, the ISO 7743 standard defines the geometry of the specimens and the testing conditions, while for stiffer materials ISO 604 is the reference to be followed. To analyse the hydrostatic component several different methods are available, all based on the application of an hydrostatic pressure in order to cause a change in volume, being the proportionality factor the bulk modulus (K). For a stiffer material it is possible to find a relation between E and K from the linear elasticity theory, while for softer materials, the neo-Hookean law gives a satisfactory relationship between both properties [8].

3 Techniques for the Mechanical Characterization of Adhesives Under Shear

A well-designed joint should always avoid peel and cleavage stresses in the adhesive, and thus the joint should mainly encounter shear loads when in service. Therefore, the shear properties of the adhesive must be accurately determined to allow for a successful joint design. The adhesive shear modulus (G) can be related to the E obtained in the uniaxial tensile test with the linear elasticity theory, as follows:

$$G = \frac{E}{2(1 + \nu)} \tag{1}$$

where ν represents the Poisson's ratio of the adhesive material.

However, this approach to obtain the shear properties of the adhesive is not the most reliable as the existence of defects in the tensile specimens has a considerable influence in the obtained results. In tension, the strain to failure of the specimens is highly dependent on the presence of defects, whereas in shear the presence of certain defects does not significantly affect the mechanical behaviour, as the remaining area is capable of withstanding further deformation. Thus, the shear modulus, shear strength to failure and shear deformation to failure should all be determined from the tests listed in the following subsections [1, 8].

3.1 Lap Joints

Lap joint tests are still one of the most used tests in an industrial context, given their simplicity and very low cost. Generally, single lap joints are used for this purpose and they are described in ASTM D 1002 and ISO 4587 standards. The use of a single lap joint allows to measure the apparent shear strength of an adhesive (τ_{av}), which can be obtained by dividing the maximum load (P_{max}) by the bonded area (A), as follows:

$$\tau_{av} = \frac{P_{max}}{A} \tag{2}$$

This shear strength is referred to as apparent because the real state of stress of a single-lap joint is very complex as a result of existence of peel stress in the adhesive, given the misalignment of the joint when loaded. This is schematically represented in Fig. 2. In addition, joint strength is highly dependent on the existence of a fillet at the ends of the overlap, which is known to increase the joint strength [13–15].

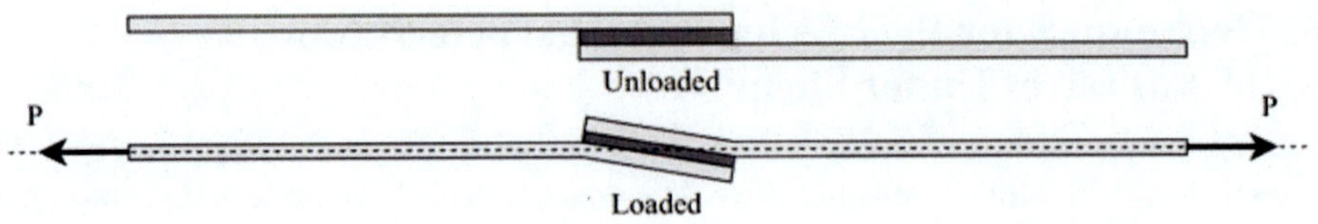

Fig. 2 Schematic representation of a single-lap joint, mounting and loaded behaviour

3.2 *Iosipescu (V-Notched Beam) and Arcan (V-Notched Plate)*

Proposed by Iosipescu [16], the V-notched beam method can test both bulk and joint samples and is standardised by ASTM D5379. The V-notched beam specimen is a short beam loaded in a way to avoid bending moment at the notch region in order to obtain a pure shear stress at the notch. A four-point bending apparatus with two V notches at its centre, as represented in Fig. 3, can be used for this purpose. The V-notched plate method was developed with the purpose of analysing the shear behaviour of the materials, mostly polymers. The Arcan specimen has two V notches as the Iosipescu specimen, but a plate is used instead of a beam. Therefore, this procedure is very versatile, allowing to test specimens under Mode I, Mode II and mixed-mode fracture behaviour by using a simple fixture with multiple holes for pinning the specimen into a machine, as represented in Fig. 3. The measurement principle of the shear strain and shear stress is the same as for Iosipescu method, Fig. 3, with similar stress concentration factors at the notch roots.

Strain gauges are mounted at an angle of −45° and + 45° with the specimen axis in order to measure the shear strain. Considering P as the applied force, w the specimen thickness and t the distance between the notch roots, the shear strain (γ) is given by:

$$\gamma = \frac{\delta_1 - \delta_2}{t} \tag{3}$$

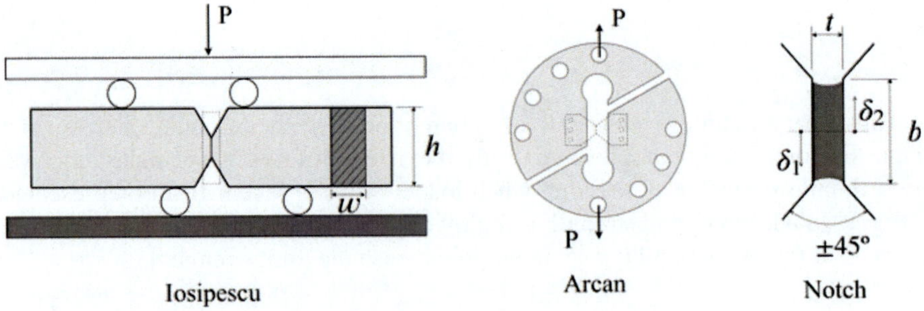

Fig. 3 Asymmetric four-point bending test, Arcan test and notch detail

where $\delta_1 - \delta_2$ represent the relative vertical motion of two points on either side of the center line of the specimen (see Fig. 3) [8].

For the shear stress (τ):

$$\tau = \frac{P}{wt} \tag{4}$$

Considering the strain gauges deformations at an angle of +45° with the specimen axis, $\varepsilon_{45°}$, and with an angle of −45° with the specimen axis, $\varepsilon_{-45°}$, the shear modulus is given by [1]:

$$G = \frac{P}{wt \cdot (\varepsilon_{45} - \varepsilon_{-45})} \tag{5}$$

The V-notched beam and V-notched plate tests are still widely used procedures, especially for composite materials and adhesives, as shown by Stojcevski et al. [17]. Nevertheless these methods still create stress concentrations at the notch roots that will affect the obtained results, as studied by Adams and Walrath [18, 19] and Kumosa and Hull [20]. Machining the notch roots with a certain radius of curvature will decrease the stress concentration factor [21]. A finite element analysis is also suggested in order to determine the concentration factor for a given geometry allowing to correct the experimentally obtained results, even tough, the shear stress calculated within the cross-sectional area is a good approximation [1].

Grabovac and Morris [22] obtained better results (an increase on the shear strength to failure value) with V-shaped beam tests than with lap shear joints specimens, as a result of the much lower stress concentration factor of this test. Regarding the repeatability of the process, Grabovac and Morris [22] obtained a standard deviation for the shear strength of tougher adhesives of less than 10%, while for the modulus measurements standard deviations of 15–20% were obtained, which means that this test allows for accurate results shear but it is still highly dependent on the manufacturing process of the specimens.

Créac'hcadec et al. [23] designed a stress concentration-free bonded Arcan tensile compression shear test specimen (Arcan TCS) for the evaluation of the adhesive mechanical properties, for a specific epoxy-based adhesive. However, the Arcan TCS was not found to be very practical, as it must be modified to test different adhesives in order to obtain minimal stress-concentration factors at the notches.

3.3 *Thick Adherend Shear Test (TAST)*

The thick adherend shear test (TAST) avoids the complex state of stress existent on single lap joints by employing a thicker substrate, Fig. 4, much stiffer than the adhesive, which avoids bending moments and creates a uniform shear stress distribution

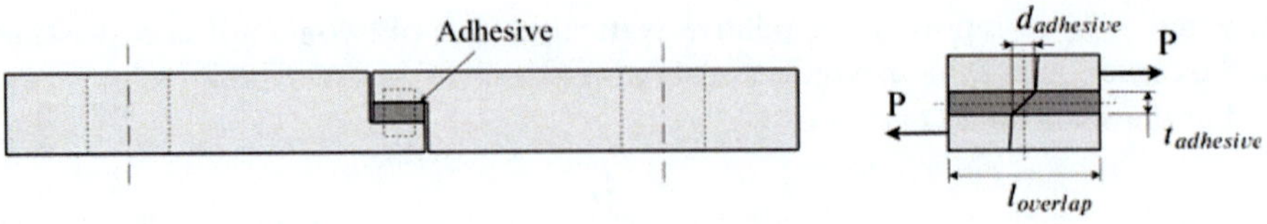

Fig. 4 Thick adherend shear test, TAST, schematic representation

over most of the overlap. The specimens for this test are standardised by ASTM D 3983 and ISO 11003-2, each with a different geometrical configuration.

By considering the overlap length of the adhesive, as represented in Fig. 4, and assuming an uniform shear stress over the entire length of the adhesive, $l_{overlap}$, the shear strain (γ) and τ are defined by:

$$\gamma = \frac{d_{adhesive}}{t_{adhesive}} \tag{6}$$

$$\tau = \frac{P}{l_{overlap} \cdot b} \tag{7}$$

where the adhesive thickness is $t_{adhesive}$ and b the width of the adhesive. The adhesive deformation is represented by the linear displacement at the interface between the adhesive and the adherend, $d_{adhesive}$. Despite the approximately uniform shear stress that is obtained along the overlap, both experimental and numerical tests have shown that stress concentration and peel stresses still appear at the ends of the overlap. Taking this into consideration, when testing stiffer adhesives it is highly recommendable to use a 45° fillet in order to minimize stress and strain concentrations at the end of the overlap [24]. Cognard et al. [25] presented a modified TAST test to reduce edge effects of the joint by machining the substrate and the adhesive in order to obtain a geometry that will avoid stress concentrations at that location.

3.4 Butt Joints in Torsion

Shear properties of adhesives measured in torsion generally offer better experimental results than other shear tests. When a joint is subjected to pure shear, the entirety of the bonding area will contribute to the joint strength. A defect such as a void, if present on a bulk tensile specimen, will almost always act as a stress riser and crack initiation point, ultimately leading to an early failure of the specimen. In this case, the experimentally determined failure strain and stress will not correspond to the strain failure of the material. However, when a joint is tested under shear conditions the existence of a void does not significantly affect the joint, as all the area of the bondline is being loaded and contributing to the shear resistance of the joint. Considering that torsion testing creates a pure shear stress on the specimen, it is expectable to obtain

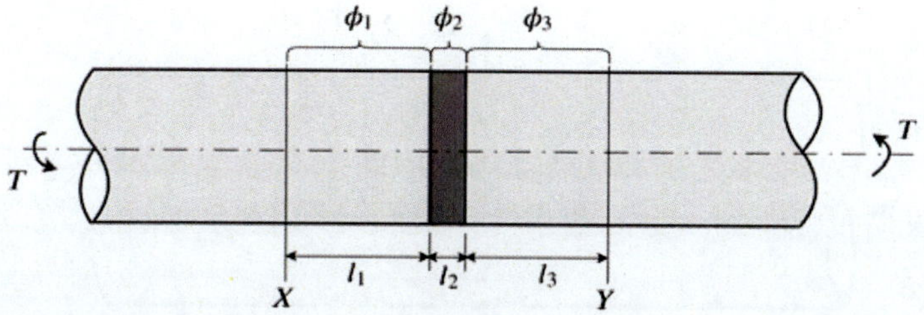

Fig. 5 Solid butt joint for torsion test. Adapted from [26]

a pure shear state of stress on the adhesive, thus leading to more realistic shear properties of the adhesive.

3.4.1 Solid Butt Joint

Considering a butt joint, as represented in Fig. 5, Adams et al. [26] obtained a relationship between the rotation ϕ measured at two different points X and Y at a distance $l = l_1 + l_2 + l_3$, where $\phi_i = f(G_i, l_i, r_i)$ for each material i, were G_i is the shear modulus of the material i and l_i, r_i the length and radius of the section i, respectively.

From the relationship between the overall rotation ϕ_{ov} and the rotation of each section (obtained from the material properties and geometry), $\phi_{ov} = \phi_1 + \phi_2 + \phi_3$ it is possible to conclude that if a testing machine is capable of measuring the rotation of a butt joint specimen in at least for two different points, then the stress field in the adhesive can be deduced.

3.4.2 Torsion of a Circular Cross-Section

The shear stress τ on a cross-section of the adhesive is a function of the shear strain γ and the section material:

$$\tau = G \cdot \gamma \tag{8}$$

Under torsion two cross-sections, at a distance l, will rotate with respect to each other through an angle ϕ, which is proportional to the length l [27]:

$$\phi = \theta \cdot l \tag{9}$$

where θ represents the angular twist of the unit-length section. For an applied torque (T) in the elastic domain, τ is given by:

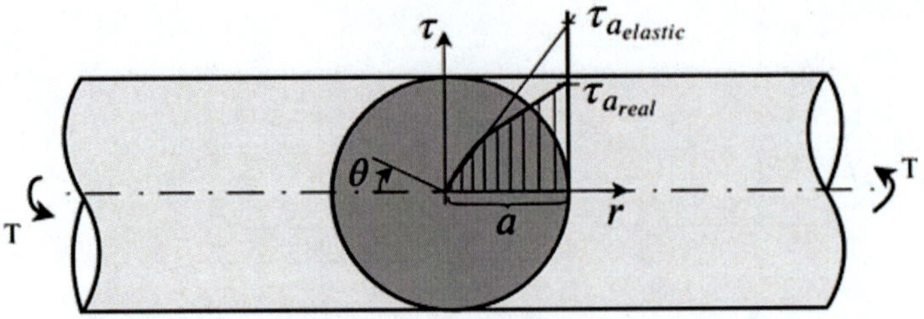

Fig. 6 Shear stress on a cross-section of the adhesive

$$\tau = \frac{T}{J} \cdot r \tag{10}$$

Considering the plasticity of the adhesive, it is not possible to assume that the τ in the adhesive layer varies linearly with the radius (r) of the section. In Fig. 6 the maximum shear stress obtained in the adhesive is represented, assuming linear elastic deformation ($\tau_{a_{elastic}}$) and considering the plasticity of the material ($\tau_{a_{real}}$).

Nadai and Wahl [27] have studied the non-linear stress–strain relationship for circular sections and proposed a method to determine the unknown stress–strain curve of the material for pure shear from a moment-twist curve obtained with a torsion test. The twisting moment (T) at a cross-section is given by:

$$T = 2\pi \int_0^a \tau \cdot r^2 dr \tag{11}$$

Since $\gamma = \theta \cdot r \Rightarrow d\gamma = \theta dr$, $\gamma_a = a \cdot \theta$ and $\tau = f(\gamma)$, Eq. 11 can be rewritten as:

$$T \cdot \theta^3 = 2\pi \int_0^{\gamma_a} f(\gamma) \cdot \gamma^2 d\gamma \tag{12}$$

Therefore, we have:

$$\frac{d}{d\theta}\left(M\theta^3\right) = 2\pi \cdot f(a\theta) \cdot a^3\theta^2 = 2\pi a^3 \cdot \theta^2 \cdot \tau_a \tag{13}$$

where τ_a corresponds to the shear stress when $r = a$. Considering the Eq. 10 and that the second polar moment of the circular section is $J = \frac{\pi}{2} \cdot r^4$, we have for τ_a [27]:

$$\tau_a = \frac{1}{2\pi a^3}\left(\theta \cdot \frac{dT}{d\theta} + 3T\right) \tag{14}$$

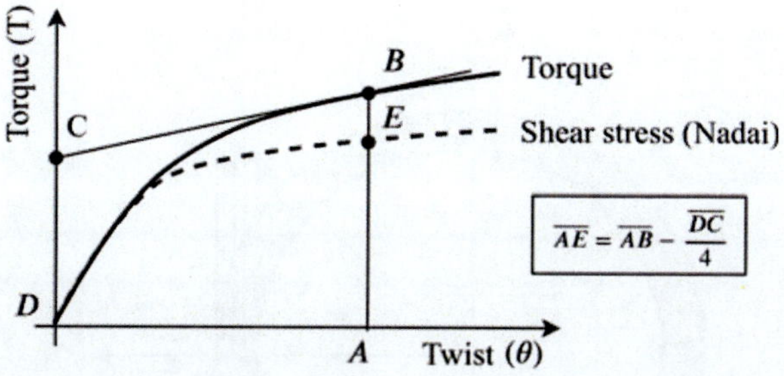

Fig. 7 Nadai's correction Adapted from [8]

Equation 14 allows to obtain the shear stress–strain curve of the adhesive from a torsion test, considering its plasticity.

In Fig. 7 the shear stress obtained from a torque-twist curve is represented for a generic point. Note that the shear stress (point E) can be obtained by reducing the vertical distance of the point B by a quarter of the vertical distance of the point C given by the interception of a curve tangent to point B with the vertical axis [27].

3.4.3 Napkin-Ring Test

An alternative to the butt joint test is the Napkin Ring test, used to determine the shear stress–strain curve as well. By considering a tubular setup, as represented in Fig. 8, it is possible to determine G and the shear strength of the adhesive.

By having a torque T applied at the two opposites ends of the tube, the maximum shear stress (τ_{max}) at the outer radius (r_o) is given by:

$$\tau_{max} = \frac{T \cdot r_0}{J} \tag{15}$$

where the second polar moment of the ring is given by:

$$J = \frac{\pi}{2}\left(r_o^4 - r_i^4\right) \tag{16}$$

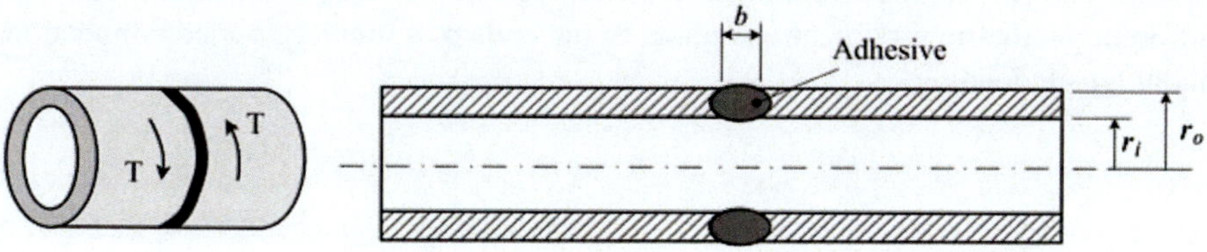

Fig. 8 Napkin ring test, schematic representation. Adapted from [8]

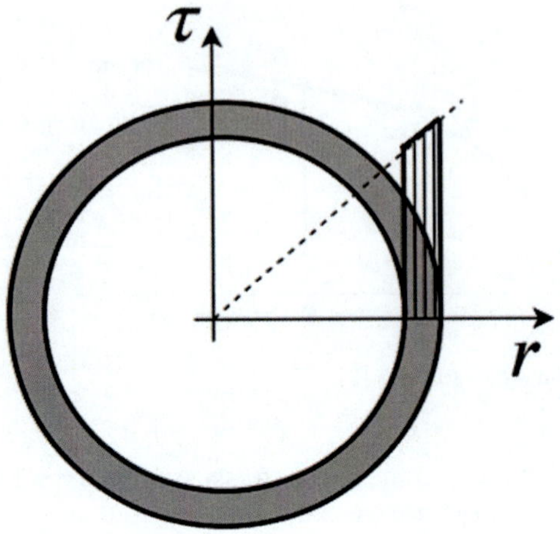

Fig. 9 Shear stress on a cross-section of the adhesive

The shear modulus is given by:

$$G = \frac{b}{J}\left(\frac{T}{\phi}\right) \tag{17}$$

where (T/ϕ) is obtained from a torque-twist curve.

For this test, the shear stress in the adhesive can be considered to be constant and given by the maximum value obtained with Eq. 15, since the difference in the shear strength of the adhesive in the outer radius (r_o) and the inner radius (r_i) is reduced, as represented in Fig. 9. If $r_i \approx r_o$, then τ will be approximately the same.

In theory, the Napkin-Ring test should easily deliver better results as the shear stress can be considered as constant and there is no need to consider the plasticity of the adhesive, as is the case for the solid butt joint. However, the manufacture of the specimens is much more complex in this case, as misalignments of the two components can easily occur during the bonding process, especially when working with low viscosity adhesives. Adams and Coppendale [28] and Chai [7] have shown that the fillets that appear in the bonding zone will strongly influence the results obtained under different load conditions, thus making the calculation of the shear strength and G less exact. Similar conclusions were obtained by Ferreira et al. [29] for different geometric modifications made to the ends of a tubular bonded connection under tensile loading.

4 Overview of Torsion Testing Equipment Currently Used to Characterise Adhesives

Currently, there is large number of commercial manufacturers of torsion machines. However, most of these manufacturers are not specialised in the design of torsion machines for adhesive bonding applications, offering instead equipment with more universal testing capabilities. Even though some of these machines are capable of perform torsion tests on adhesives, they are generally better suited to test other materials, such as composites, steel, aluminium alloys and polymers [30–32]. In general, commercially available machines can deliver torque at different ranges, starting from 5 N m until 2000 N m. For testing polymers, torsion machines have a maximum torque of around 250 N m, while for testing metals higher torque values are necessary. In order to test adhesives, the referred machines need only to provide a maximum torque of 60 N m while still being capable test at much lower torque values for the characterization of flexible adhesives. Another characteristic of these machines is that the commercially available models ensure high axial stiffness of the system, but do not provide methods for precise alignment of the specimen. Therefore, problems of concentricity between one side of the specimen and the other will appear meaning that radial and/or axial forces or even displacements that are not absorbed will have an influence on the testing results.

There are several patents describing devices for measuring the shear strength of materials, determining the shear stress through the application of a torque to a specimen. Matsumoto et al. [33] presented a mechanism capable of performing fatigue tests in a torsional testing machine. However, this patent does not describe the ability of the device to absorb non shear loads. Matsumoto et al. [34] also describes a device that promotes torsional tests, but as in the previous patent, it does not describe any kind of mechanism to remove deformations or stresses besides the ones introduced by the applied torque. That document focuses on the ability to perform both static and dynamic tests in a single device. Lastly, there are two patent documents by Thomas et al. [35] and Fauner [36], that refer devices for measuring the shear strength of structural adhesives. However, once again, these documents do not make any reference to a kinematic chain with the ability to absorb residual stresses when torque is applied.

5 Description of the Novel Torsion Machine

This section describes the novel machine designed for torsion tests of adhesives. As a result of the drive-line solution proposed, this equipment is able to apply a pure shear loading on the specimen, which results in more accurate results. The torsion testing machine uses torsionally rigid couplings, self-centering bearings and a car able to freely move in the vertical direction, enabling the adjustment of the machine during the tests, compensating any undesirable displacement, and thus removing any spurious load on the specimen.

The centerline is in the vertical direction (perpendicular to the floor) to eliminate the negative effects of the gravitational acceleration. In addition, the machine has a construction that compensates the vertical forces. The components are able to compensate assembly or machining errors, as well as deformations in the specimen itself. The use of different sized specimen is also possible thanks to the car mobility and the claws. For the mechanical action, an assembly of a servomotor, drive and a planetary gearbox was made. The angular displacement is measured from the encoder signals that are supplied by the servomotor drive. For increased accuracy, a measuring device, applied directly to the specimen, may be used. All those components are electrically powered so it was necessary to conceive a control cabinet to power the components, to condition the signal and to provide overload protection.

Because this test apparatus is capable of controlling the torque and angular displacement that is applied to the specimen, it can also perform creep tests, torque rate control tests, relaxation tests and displacement rate control tests. The machine is schematically represented in Fig. 10 and its main components are listed.

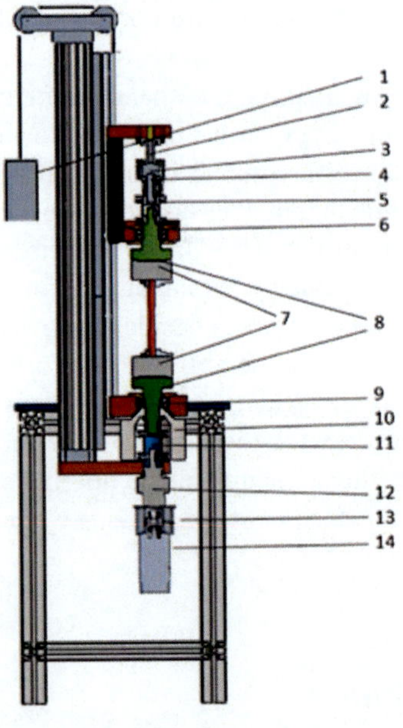

Components	
1, 10	Counterweight
2	Torque transducer
3	Clutch
4, 8	Shaft
5, 11	Torsionally rigid coupling
6, 9	Ball bearings
7	Claws
12	Reducer gearbox
13	Coupling
14	Servomotor

Fig. 10 Novel torsion machine, schematic representation

5.1 Main Specifications

The specifications of the torsion machine were defined to match the needs associated with testing adhesive joints, including the possibility of testing with different geometries, such as bulk specimens, butt joints and tubular joints.

- Maximum torque: 60 N m;
- Tests speed: 0.05–1 rpm (0.05–0.105 rad/s);
- Resolution of the angular transducer: 0.0005°;
- Spurious tensile stresses below 1 MPa;
- Dimensions of the specimen allowed
  - Diameter: 5–30 mm;
  - Length: 50–150 mm.

Usually, machines for torsional testing have horizontal axis stands with certain components (such as bearings and rigid couplings) to compensate for misalignments of the kinematic chain that lead to the appearance of residual stresses when loaded horizontally due to the weight of the machine components. To solve this problem, it was decided to align the construction solution vertically, as shown in Fig. 11. In order to make the device more versatile, a mobile car which carries the upper assembly was designed, enabling the adjustment of the distance between claws, making the machine highly adaptable to different sized specimens.

5.1.1 Claws and Supports

In the selection of the claws, certain important conditions were taken into consideration, such as the maximum and minimum spacing of the grips and their ability to prevent the sliding of the specimen, ensuring effective transfer of the torque load. Grip claws are particularly suitable for this purpose since they have high structural rigidity. The gripping area can either have a square section or a circular section, being necessary to use a claw with four tightening grips. The selected claw configuration avoids relative sliding between the grips and the specimen upon the application of a torque, thereby preventing errors when measuring the angular deformation of the adhesive. This claw also has the characteristic of being self-centering, thus ensuring undesirable spurious loadings. Two supports are placed in the centre of the kinematic chain, holding the claws in place. Nevertheless, to avoid possible misalignment between the extremities of the specimen, the supports must allow for more than one degree of freedom between the claws and the torsionally rigid coupling, as shown in Fig. 12. Self-centering ball bearings with low friction were chosen for this purpose. This kind of bearing allows for oscillations of approximately 4° from the central position. Note that the shaft between the support of the claws and the bearing has a section reduction with a large fillet to avoid stress concentrations.

Fig. 11 Picture of the torsion machine

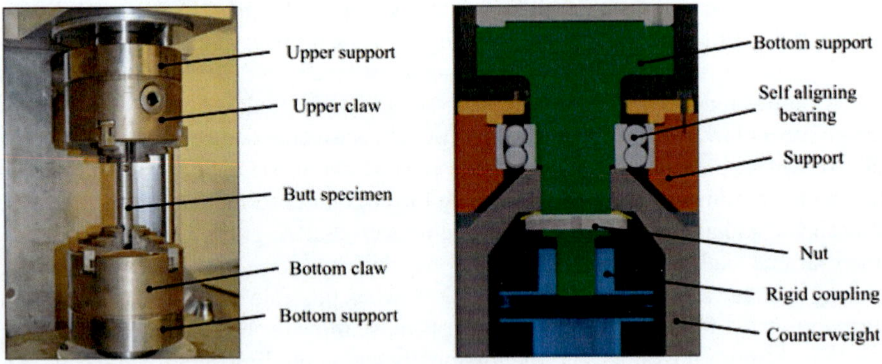

Fig. 12 Claws and supports representation

5.1.2 Claw's Counterweight

With the machine being positioned in the vertical position, it is mandatory that the claw's centre of mass, along with that of the linking shaft, are situated below the bearings. A counterweight was thus included at the bottom of the kinematic chain, as shown in Fig. 12. This prevents the lower claw from becoming an inverted pendulum with unstable behaviour.

5.1.3 Mobile Elements

As mentioned above, the kinematic chain elements, from the servomotor to the lower claw, are all fixed to a vertical column, and do not have the freedom to move axially. However, all elements from the upper claw to the torque transducer, are fixed to a mobile car, which is free to move axially in linear guides and bearings, which are directly tightened to the vertical column. Using this concept, no axial loads are transferred to the specimen being tested. To keep the mobile car in a static position, another counterweight was placed in the rear of the testing machine, as shown in Fig. 13. This solution was numerically simulated to determine the load

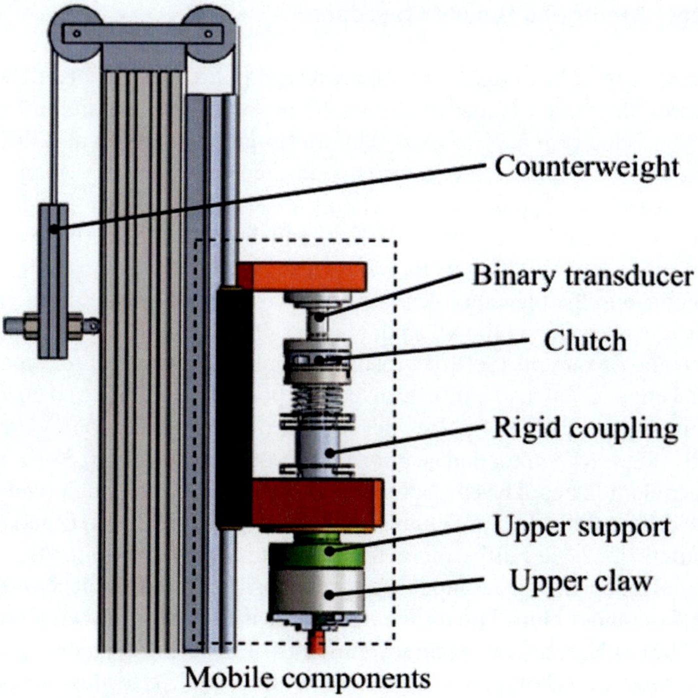

Fig. 13 Counterweight and mobile elements

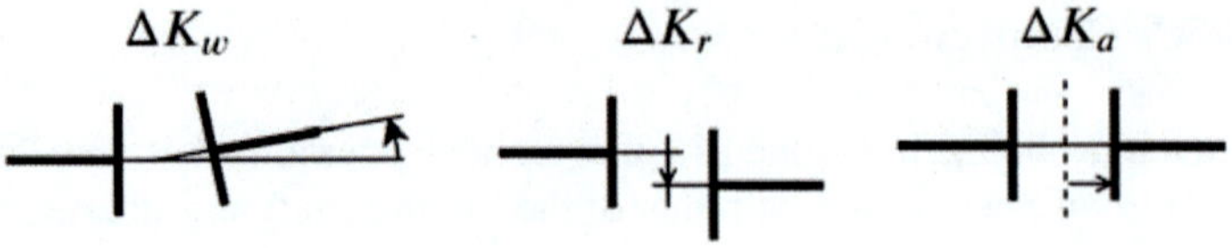

Fig. 14 Types of misalignments

absorption capacity which shields the adhesive layer from any spurious loads, created by deformations on the machine itself.

5.1.4 Couplings

The torsionally rigid couplings in the machine allow to avoid various types of misalignments, shown in Fig. 14, that result from centering errors of the assembly and construction of the table. Without these couplings, the self-centering bearings would have no practical effect. The direct connection of the linking shaft to the transducer or to the gearbox would be rigid, preventing the desired effect.

5.1.5 Servo Motor and Binary Transducer

The servomotor is an AC brushless motor with ten poles and a resolver (two poles) integrated into the motor. It features a locking brake and the maximum continuous torque at low velocity is 2 N m. A maximum torque of 6.6 N m at 2300 rpm can be attained at 230 V. This servomotor is connected to a planetary gearbox with a reduction ratio of 216:1, maximum torque of 80 N m and an overall yield of 78%. The maximum velocity is 10.7 rpm = 2300/21 and the torque in continuous operation is 337 N m = 2 × 216 × 0.78 (this torque is electronically limited to 70 N m in order to avoid damage to the planetary gearbox).

The drive is equipped with two analog inputs (14 bits and 10 bits; ±10 V differential), two analog outputs (±10 V), resolver input and one output for emulating an incremental encoder. By using this resolver, it is possible to emulate an encoder with a maximum signal of 16,384 pulses per revolution. The resolution of the encoder angular position (servomotor and gearbox) is 0.000025° = 360°/(16,384 × 4 × 216). The data acquisition board has a resolution of 12-bit and sixteen single-ended analog inputs or 8 differentials. Its maximum sampling frequency is 200 kHz and it has two analog outputs of 12-bit. Furthermore, this board has a digital 24-bits I/O, which can be configured independently as input or output. It also has three 16-bit counters thanks to the 82C54 counter chip. The torque transducer is non-rotating, with a maximum torque of 20,000 N m and an accuracy class of 0.1. There is a control cabinet that provides power to all components, conditions the signal, provides protection and power cut, shown in Fig. 15.

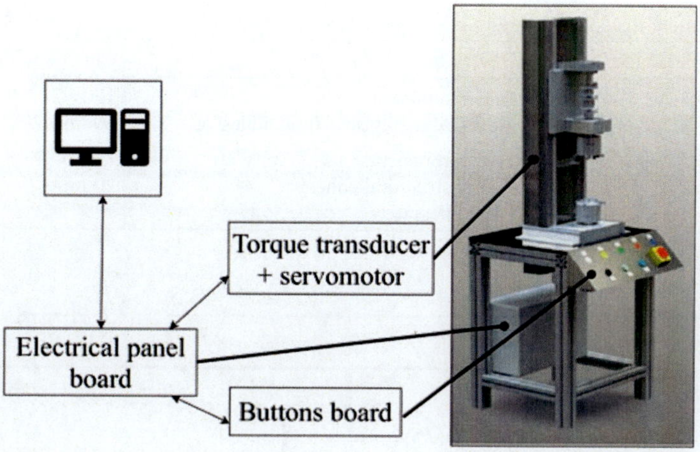

Fig. 15 Machine control system

6 Validation

A validation procedure of the testing machine was carried in the present work, using experimental test results from different structural adhesives. These results were analysed in order to characterise the materials' behaviour under shear loads performed with this novel torsion machine and with a commercial machine (tensile test machine). For the novel torsion machine, butt joints with solid adherends were used while Thick Adherend Shear Tests (TAST) were used for tests performed on the tensile testing machine. The ultimate objective of this procedure is to compare the differences in the measured mechanical response of the adhesives, using these two distinct approaches.

6.1 Experimental Details

With the aim of validating the torsion machine for shear characterization of different adhesives, two distinct adhesives were used, an epoxy-based adhesive and a modified epoxy adhesive, expected to have a more ductile behaviour, Table 1.

TAST specimens were tested according to ISO-110032 standard, as represented in Fig. 16. The specimen (thus, the adherend) was machined from DIN St37 low-strength steel. The TAST specimens were manufactured using a mould, as shown in Fig. 17. During the assembly of the substrates, the adhesive overlap is ensured using shims, which are previously coated with a mould releasing agent to allow for their posterior removal. The specimens were sandblasted and degreased with acetone in order to create a surface suitable for good adhesion of the adhesive. The film adhesive was cut to the precise dimensions of the overlap and carefully placed in position,

Table 1 Structural adhesives used in the experiment

Name (Manufacturer)	Description	Cure condition
AF 163-2K (3M)	One component Modified epoxy structural adhesive	130 °C 60 min
Araldite 420 A+B (Huntsman)	Two components Epoxy structural adhesive Past adhesive	120 °C 20 min

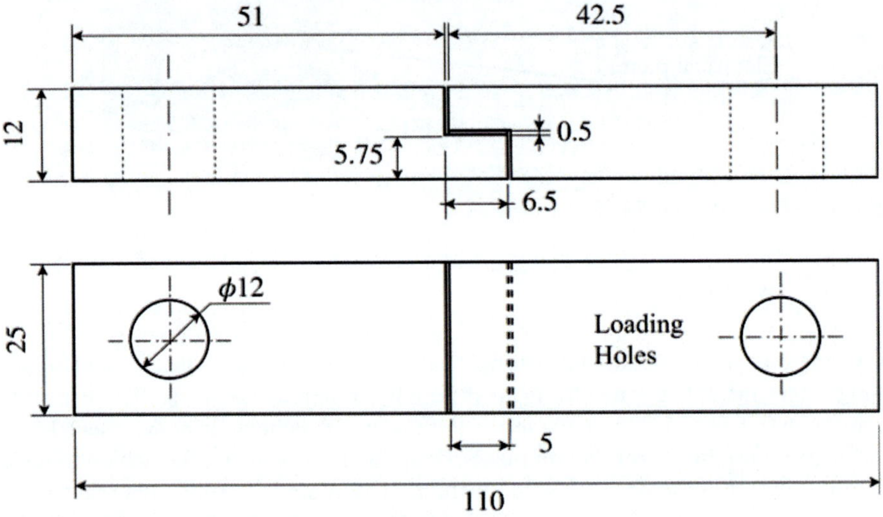

Fig. 16 TAST specimen geometry, dimensions in mm

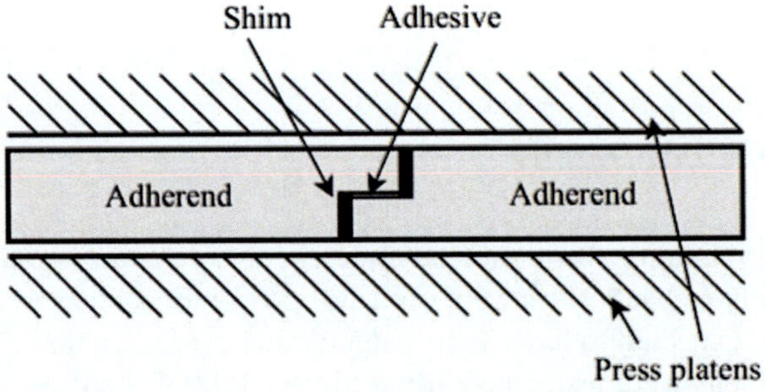

Fig. 17 Schematic view of the mould for TAST specimens. Adapted from [8]

while the paste adhesive was applied using a spatula. The two adhesives have been cured according to the manufacturer recommendations, presented in Table 1. Before testing the specimens, the spacer shims and the excess adhesive were removed.

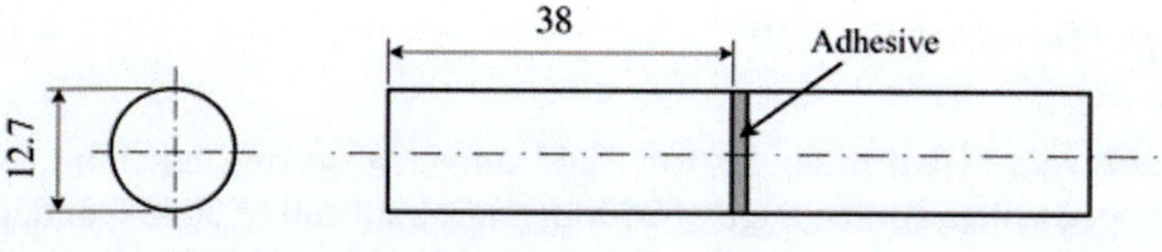

Fig. 18 Butt specimen geometry, dimensions in mm

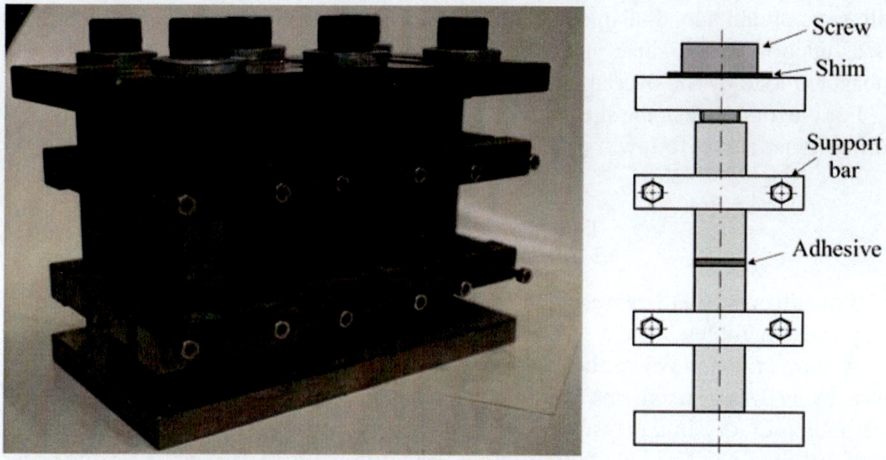

(a) Picture of the mould (b) Mounting

Fig. 19 Picture of the mould and schematic representation of the mounting system, adhesive thickness controlled by the use of shims

The butt joints with solid adherends for torsion testing were manufactured according to ASTM D 2095 standard, with the dimensions shown in Fig. 18. The specimens were manufactured using a special mould, shown in Fig. 19, proposed by da Silva et al. [8]. The mould is capable of simultaneously controlling the alignment of the specimens' substrates and the adhesive's layer thickness using shims located on top of the mould, as represented schematically in Fig. 19. The substrates were sand-blasted and then degreased with acetone, to remove any particle that could potentially influence the adhesion between the adhesive and the substrate. The adhesive was then applied to the bottom substrate and the upper substrate positioned and lightly fixed by the support bars. To bring the upper substrate in contact with the adhesive, it was vertically adjusted by turning the screw on top of the mould. The substrate is free to move until it reaches the shims, which defines the adhesive layer thickness. After the curing process was completed the specimens were carefully removed from the mould and the excess adhesive removed with sandpaper.

6.2 Experimental Results

TAST joints were tested in an Instron 3367 universal testing machine (Norwood, USA) equipped with a strain gauge of 25 mm and a load cell of 30 kN, under ambient conditions (room temperature of 23 °C, relative humidity of 55%). A displacement rate of 1 mm/min was used. For the torsion tests, the presented novel torsion machine was used under ambient conditions at a test speed of 0.05 rad/s. The TAST procedure allows to obtain a load–displacement curve for the specimens, which can be plotted as a function of the shear stress and shear strain in the adhesive by dividing the measured load by the overlap area of the specimen.

For the torsion test, the shear stress–strain curve is obtained from an experimental torque-twist curve, following the application of Nadai's correction, as follows:

$$\tau_a = \frac{1}{2\pi a^3}\left(\theta \cdot \frac{dT}{d\theta} + 3T\right) \tag{14}$$

Typical curves are represented in Figs. 20 and 21 and the main mechanical properties in Tables 2 and 3.

A comparison between the main mechanical properties obtained for both adhesives using TAST specimens and torsion specimens shows that for small strains both methods provide similar results. For the modified-epoxy (AF163-2K) the Young's modulus obtained from the two different tests differs by 15.1%, while for the shear strength the difference is only 1.6%. Thus, it can be stated that the adhesive is well characterised in shear by both methods if small strain values are considered. For the epoxy adhesive (Araldite 420 A+B) the difference found either for the Young's modulus or for the shear strength has been always lower than 5%. However, a considerable difference in the yield strength value was found for the most ductile adhesive. Figure 20 shows that significant yielding starts to appear below 10% of shear strain for the TAST specimen while for the butt joint under torsion the same value was

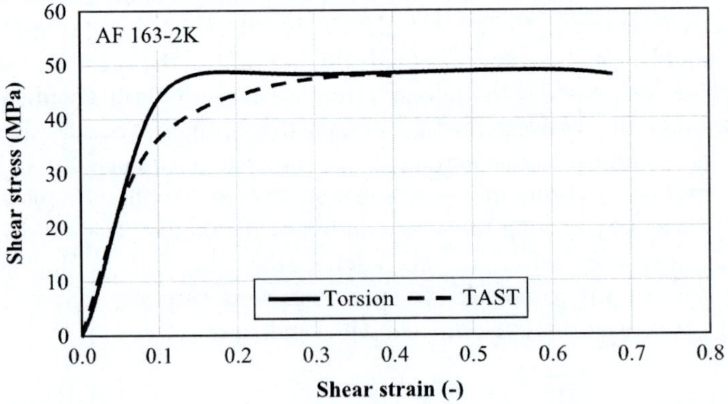

Fig. 20 Shear stress–strain curves for TAST and torsion tests, AF 163-2K

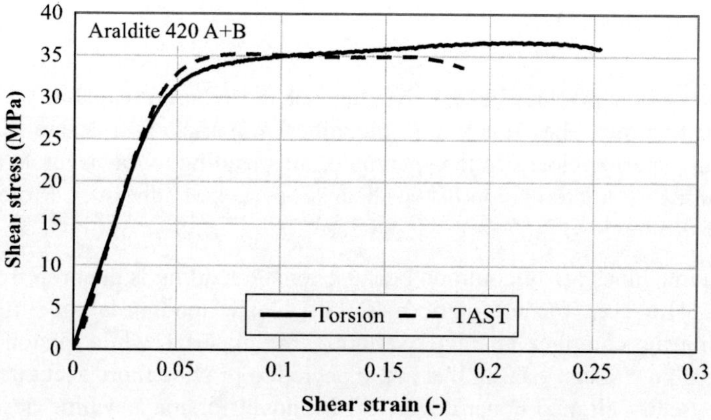

Fig. 21 Shear stress–strain curves for TAST and torsion tests, Araldite 420 A+B

Table 2 Main mechanical properties experimentally obtained for AF163-2K

	Shear modulus (MPa)	Shear strength (MPa)	Shear strain to failure (%)
TAST	443.7 ± 88.9	46.9 ± 2.6	40.8 ± 9.6
Torsion	522.5 ± 41.7	47.5 ± 1.9	69.1 ± 5.9

Table 3 Main mechanical properties experimentally obtained for Araldite 420 A+B

	Shear modulus (MPa)	Shear strength (MPa)	Shear strain to failure (%)
TAST	776.5 ± 38.1	34.9 ± 2.1	18.7 ± 8.9
Torsion	756.9 ± 47.9	33.7 ± 2.7	25.3 ± 6.8

around 10%. The main reason for this to happen is the fact that the TAST specimens have a significant stress concentration factor at the ends of the overlap, which causes the adhesive to yield much earlier, while for the torsion specimens the adhesive is under a stress state very close to pure shear.

The most significant difference identified on all tested specimens was the shear strain to failure value. For AF163-2K the obtained shear strain to failure was 70% higher when obtained with the butt joint in torsion than with a TAST specimen, while for Araldite 420 A+B the difference is 35%. This improvement clearly reflects the ability of the torsion tests to characterize the adhesive under shear, especially for the purpose of determining the shear strain to failure of the material. Lastly, torsion tests presented good repeatability and accuracy when compared with TAST specimens, when considering the three tests performed for each adhesive.

7 Conclusion

This novel torsion machine takes advantage of the inherent capability of torsion tests to create a pure shear loading in the adhesive by having a design focused on avoiding any spurious loads in the specimen, thus leading to the accurate results in the mechanical characterization of the shear properties of adhesives. More detailed conclusions follow:

- Torsion machines are uncommon because tensile loading is the preferred testing method. However, for polymeric materials, tensile loading is not sufficient for achieving the complete characterisation of the material, while torsion loadings provide a pure shear loading that are expected to provide more accurate results;
- The vertically aligned construction of the novel torsion machine described in this work avoids unwanted loads and spurious bending moments acting on the specimen during the torsion test procedure;
- Moreover, the mechanisms of the machine enable its constant adjustment during the test, compensating any undesirable displacement, thus removing non torsional loads from the specimen;
- The presented torsion machine is able to completely characterise an adhesive material and its performance during the torsion loading;
- The determination of the strain to failure is properly characterised when used this novel torsion machine.

Acknowledgments The authors would like to thank the Portuguese Foundation for Science and Technology (FCT) for supporting the individual grant SFRH/BPD/96992/2013.

References

1. da Silva LFM, Öchsner A, Adams R (2018) Handbook of adhesion technology
2. Dolev G, Ishai O (1981) Mechanical characterization of adhesive layer in-situ and as bulk material. J Adhes 12:283–294
3. Lilleheden L (1994) Mechanical properties of adhesives in situ and in bulk. Int J Adhes Adhes 14:31–37
4. da Silva LFM, Adams RD (2005) Measurement of the mechanical properties of structural adhesives in tension and shear over a wide range of temperatures. J Adhes Sci Technol 19:109–141
5. Dean GD, Duncan B, Adams RD, Thomas RD, Vaughn LF (1996) An evaluation of bulk and joint specimen tests for determining the shear properties of adhesives
6. Bryant RW, Dukes WA (1965) The measurement of the shear strength of adhesive joints in torsion. Br J Appl Phys 16:101–108
7. Chai H (2004) The effects of bond thickness, rate and temperature on the deformation and fracture of structural adhesives under shear loading. Int J Fract 130:497–515
8. da Silva LFM, Dillard D, Blackman B, Adams R (2012) Testing adhesive joints, best practices
9. Banea M, da Silva LFM, Campilho R (2012) Moulds design for adhesive bulk and joint specimens manufacturing. Assem Autom 32:284–292

10. da Silva LFM, Adams RD, Gibbs M (2004) Manufacture of adhesive joints and bulk specimens with high-temperature adhesives. Int J Adhes Adhes 24:69–83
11. Banea M, da Silva LFM (2010) The effect of temperature on the mechanical properties of adhesives for the automotive industry. Proc Inst Mech Eng Part L J Mater Appl 224:51–62
12. Katnam KB, Stevenson JPJ, Stanley WF, Buggy M, Young TM (2011) Tensile strength of two-part epoxy paste adhesives: influence of mixing technique and micro-void formation. Int J Adhes Adhes 31:666–673
13. Crocombe AD, Adams RD (1981) Influence of the spew fillet and other parameters on the stress distribution in the single lap joint. J Adhes 13:141–155
14. Adams RD, Peppiatt NA (1974) Stress analysis of adhesive-bonded lap joints. J Strain Anal 9:185–196
15. Doru MO, Özel A, Akpinar S, Aydin MD (2014) Effect of the spew fillet on adhesively bonded single-lap joint subjected to tensile loading: experimental and 3-D non-linear stress analysis. J Adhes 90:195–209
16. Iosipescu N (1967) New accurate procedure for single shear testing of metals. J Mater 2:537–566
17. Stojcevski F, Hilditch T, Henderson LC (2018) A modern account of Iosipescu testing. Compos Part A Appl Sci Manuf 107:545–554
18. Adams DF, Walrath DE (1987) Further development of the losipescu shear test method. Exp Mech 27:113–119
19. Adams DF, Walrath DE (1987) Current status of the Iosipescu shear test method. J Compos Mater 21:494–507
20. Kumosa M, Hull D (1987) Mixed-mode fracture of composites using Iosipescu shear test. Int J Fract 35:83–102
21. Swanson SR, Messick M, Toombes GR (1985) Comparison of torsion tube and Iosipescu in-plane shear test results for a carbon fibre-reinforced epoxy composite. Composites 16:220–224
22. Grabovac I, Morris CEM (1991) The application of the Iosipescu shear test to structural adhesives. J Appl Polym Sci 43:2033–2042
23. Créac'hcadec R, Sohier L, Cellard C, Gineste B (2015) A stress concentration-free bonded arcan tensile compression shear test specimen for the evaluation of adhesive mechanical response. Int J Adhes Adhes 61:81–92
24. Kadioglu F, Vaughn LF, Guild FJ, Adams RD (2002) Use of the thick adherend shear test for shear stress-strain measurements of stiff and flexible adhesives. J Adhes 78:355–381
25. Cognard JY, Créac'hcadec R, Sohier L, Davies P (2008) Analysis of the nonlinear behavior of adhesives in bonded assemblies—comparison of TAST and Arcan tests. Int J Adhes Adhes 28:393–404
26. Adams RD, Adams RD, Comyn J, Wake WC, Wake WC (1997) Structural adhesive joints in engineering. Springer, Berlin
27. Nádai A, Wahl AM (1931) Plasticity: a mechanics of the plastic state of matter. McGraw-Hill Book Company Incorporated, New York
28. Adams RD, Coppendale J (1979) The stress-strain behaviour of axially-loaded butt joints. J Adhes 10:49–62
29. Ferreira LRF, Campilho RDSG, Rocha RJB, Barbosa DR (2019) Geometrical and material optimization of tensile loaded tubular adhesive joints using cohesive zone modelling. J Adhes 95:425–449
30. Wang JA, Liu KC, McCabe DE, David SA (2000) Using torsional bar testing to determine fracture toughness. Fatigue Fract Eng Mater Struct 23:917–927
31. Tan T, Ren F, Wang JJ-A, Lara-Curzio E, Agastra P, Mandell J, Bertelsen WD, LaFrance CM (2013) Investigating fracture behavior of polymer and polymeric composite materials using spiral notch torsion test. Eng Fract Mech 101:109–128
32. Keller MW, White SR, Sottos NR (2008) Torsion fatigue response of self-healing poly(dimethylsiloxane) elastomers. Polymer (Guildf) 49:3136–3145
33. Matsumoto S, Miyashita H, Murauchi K (2008) JP2008267939A

34. Matsumoto S, Miyashita H, Murauchi K, Hasegawa M, Sakagami T, Mido Y, Kawai S (2014) GB20140007136 20120926
35. Thomas B, Jane L, Geiger T (2008) US20050073103 20050303
36. Fauner G (1981) US19790026564 19790403

Development of an Elasto-plastic Meshless Technique to Analyse Bonded Structures

I. J. Sánchez-Arce, L. D. C. Ramalho, R. D. S. G. Campilho, and J. Belinha

Abstract Adhesive joints are widely used in the production of goods, mainly in the transport industry. However, their industrial applications often have non-standard complex shapes. Computer simulation, like the finite element method (FEM), is widely used for their analysis but limitations still exist. Meshless methods have been in development and offer an option to overcome some limitations of the FEM; however, these are still in development. In this work, an efficient meshless method, the 'natural neighbour radial point interpolation method' (NNRPIM), has been applied to the analysis of adhesive joints including an elasto-plastic formulation for the adhesive. First, experimental data corresponding to four overlap lengths (L_O) and two different adhesives were measured, as a benchmark. Afterwards, joint strength ($P_{\max}$) was analytically obtained as a second benchmark. Then, all the joint geometries were simulated utilising the FEM and NNRPIM methodologies, and P_{max} were calculated from those simulations. Finally, the results were compared against the first and second benchmarks. The meshless method proved to be a good alternative to the FEM, providing similar strength prediction. Moreover, the stress distribution curves were compared. In conclusion, the NNRPIM provides accurate results and could be utilised for further study of adhesively-bonded joints.

Keywords Meshless methods · Elasto-plasticity · Adhesive joints · SLJ plastic analysis · Adhesive joint strength · NNRPIM · Adhesive plasticity · Exponent Drucker-Prager

---

I. J. Sánchez-Arce (✉) · L. D. C. Ramalho · R. D. S. G. Campilho · J. Belinha
INEGI, Instituto de Ciência e Inovação em Engenharia
Mecânica e Engenharia Industrial, Porto, Portugal
e-mail: isidrodjsa@gmail.com

R. D. S. G. Campilho · J. Belinha
Department of Mechanical Engineering, Instituto Superior de Engenharia do Porto (ISEP), Instituto Politécnico de Porto, Porto, Portugal

Lucas F. M. da Silva et al. (eds.), *Industrial Applications of Adhesives*,
Lecture Notes in Mechanical Engineering,
https://doi.org/10.1007/978-981-15-6767-4_4

Abbreviations and Nomenclature

FEM	Finite element method
NNRPIM	Natural neighbours radial point interpolation method
SLJ	Single-Lap joint
CZM	Cohesive zone modelling
EDP	Exponent or extended Drucker-Prager
SSPH	Symmetric smoothed particle hydrodynamics
SPH	Smoothed particle hydrodynamics
CLS	Critical longitudinal strain
RBF	Radial basis function,
P	Polynomial basis function,
UTM	Universal testing machine,
L_O	Overlap length
P_{max}	Joint strength
λ_{DP}	Material sensitivity
σ_1	Maximum principal stress
σ_u	Material's tensile strength
σ_y	Material's yield strength
ϵ_f	Material's failure strain
E	Young's modulus
ν	Poisson's ratio
G	Shear modulus
τ_y	Yield strength in shear
τ_f	Material's shear strength
γ_f	Failure shear strain
$\boldsymbol{\Omega}$	Problem domain
$\boldsymbol{N}$	Node set within the domain
$\boldsymbol{V_i}$	Influence cell
$\boldsymbol{x_I}$	Point of interest
$R(r_{Ii})$	Multiquadratics RBF
r_{Ii}	Euclidean norm between $\boldsymbol{x_I}$ and the nodes inside $\boldsymbol{V_i}$
J_2'	Second invariant of the deviatoric stress tensor
ϵ_p	Plastic strain
λ	Plastic multiplier for the Prandtl-Reuss flow rule
$\boldsymbol{a}$	Flow vector
$\boldsymbol{c}$	Fourth order constitutive tensor
b	Exponent of the Exponent Drucker-Prager model
q	Equivalent stress (von Mises)
p	Hydrostatic pressure
M_T	Total moment matrix
t_p	Substrate thickness
t_a	Adhesive layer thickness

L_T	Length between UTM grips
τ_{xy}	Shear stress in the adhesive layer
σ_{yy}	Peel stress in the adhesive layer
ϵ_{11}	Maximum principal strain in the adhesive layer

1 Introduction

Union of non-metallic materials like those utilised in aerospace, automotive, and medical industries can be achieved either via mechanical fastening or via adhesives. The latter provide versatility, low stress concentrations, and low weight in relation to mechanical fastening [1, 2]. Along the years, analytical models had been developed to predict P_{max} [3, 4]. However, as substrate and adhesive material development progresses, the application of such models may not be always suitable.

Alongside the analytical models, FEM has allowed the analysis of joints with higher complexity (both geometrical and material-wise), and so they have been utilised to study the elastic-plastic behaviour of adhesive joints with success by considering material non-linearity [5, 6] and utilising the von Mises yield criterion. However, the mechanical behaviour of adhesives can be sensitive to hydrostatic stresses [5, 7–9]. Therefore, the von Mises criterion provides good results on ductile materials with moderate toughness and may not be suitable for many adhesive types [8]. Similarly, cohesive zone modelling (CZM) has been utilised to determine the strength of adhesively-bonded joints [10–13] but this technique does not allow to inspect the stresses within the adhesive layer.

It has been suggested that adhesive yielding depends on the hydrostatic stresses [8, 14]. In consequence, models like Coulomb-Mohr and Drucker-Prager would be suitable for adhesives. The Drucker-Prager model can be linear, hyperbolic, and exponential [15]; the latter describes adhesive behaviour better [8, 14]. The Exponent or Extended Drucker-Prager (EDP) relates both 'deviatoric' and 'hydrostatic' stresses within an exponential curve. Its application to polymers goes back to the work of Raghava and colleagues [7, 16]. The EDP takes in account the relationship between the material strength under two different loading conditions. This relationship, known as material sensitivity (λ_{DP}), influences the yield surface; however, no agreement exists about how to determine its value. According to Raghava and colleagues [7], it is the relationship between compressive and tensile yield strengths. Similarly, Dean and colleagues [8, 14] also presented relationships between shear and tensile, compression and tensile, and shear and compression strengths. The values λ_{DP} range between 1.2 and 1.5 [7], 1.3 and 1.4 [17], and 1.4 and 1.6 [8] for a similar adhesive. Moreover, as ductility and toughness increase also does λ_{DP}, leading to convergence problems in the FEM solution. Özer and Öz [18] applied the EDP model from ABAQUS® to single-lap joints (SLJ) with mono and bi-adhesive systems along the overlaps [18]; those adhesives were Araldite® AV138 and 2015. Moreover, P_{max}, in a mono-adhesive joint, was determined as the force required to make the maximum principal stress (σ_1) to reach the adhesive's tensile strength (σ_u) [18], being this 7% below the experimental strength (for Araldite® 2015) [18, 19]. In the bi-adhesive systems, the differences were found to be below 12% with respect to experimental data [18]. Therefore, the 'maximum principal stress' criterion seems acceptable. This failure criterion was proposed before by Harris and Adams [5] for non-toughened adhesives, together with the 'maximum principal strain' criterion for toughened adhesives [5].

Although the FEM has won its reputation for several types of analysis, adhesive joints amongst those, it can experience drawbacks mainly in high deformation cases because, in such cases, the elements distort beyond their acceptable limits, and so causing convergence issues. Conversely, meshless methods do not have mesh, and so they are more suitable for these applications [20–23]. In general, meshless methods use either approximation functions or interpolation functions. The later also present the Kronecker's delta property (as in the FEM) simplifying the imposition of boundary conditions (both natural and essential) [21]. The NNRPIM is a recent meshless method with high accuracy. Also, it does not require a mesh at all, only a set of points or nodes describing the geometry. Besides its accuracy, it also provides smoother stress patterns in comparison to FEM, even with a coarser nodal distribution [21].

Meshless methods have been actively utilised to analyse adhesively bonded SLJs. The symmetric smoothed particle hydrodynamics (SSPH) together with CZM was used to analyse a double cantilever beam (DCB) test [24]. A design of experiments was also performed to determine the optimum CZM parameters; finally, modes I, II, and mixed were analysed. In such case, a difference below 7% was observed between numerical and experimental data of mode I. Moreover, for angles up to 50%, mode II also presented good correlation [24]. Later, an aluminium SLJ bonded with a ductile adhesive was analysed with the smoothed particle hydrodynamics (SPH) and the CZM, within the ABAQUS ® software, using an elastic-plastic material model [9]. Unfortunately, stresses were analysed at a very low joint extension ($\delta \approxeq 0.3$ mm) and showed some numerical noise, hindering its use for comparison. The radial point interpolation method (RPIM) together with the von Mises yield criteria were used by Bodjona and Lessard [25] in the analysis of bonded-bolted (and their combinations) joints, and a comparison with the FEM was made [25]. This is a mathematically detailed study but no adhesive or substrate properties were reported. Ramalho et al. [26] applied the NNRPIM and the critical longitudinal strain (CLS) criterion to determine P_{max} on sets of aluminium SLJs bonded with different adhesive systems and with various L_O. The CLS criterion offered a good alternative to predict P_{max} using linear elastic models [26]. In another work [27], several SLJs with various L_O and three adhesive systems were analysed using both continuous mechanics failure criteria and meshless models (NNRPIM). However, the failure criteria together with linear elastic material properties under-predicted P_{max}. In consequence, an elastic-plastic analysis is required in those cases. Recently, elasto-plasticity has been implemented into the NNRPIM method and has been applied to the analysis of aluminium specimens [23] providing good agreement with both experimental data and numerical data from FEM (both commercial packages and in-house programmed routines), considering the von Mises yield criterion and a mono-material model.

In this work SLJs with four different L_O and two different adhesive systems were analysed using a meshless technique, the NNRPIM. First, the elastic-plastic von Mises yield criterion implemented into NNRPIM by Moreira et. al. [28] was applied to the SLJs. Then, the EDP with exponent two was implemented into the NNRPIM meshless formulation. Afterwards, the SLJ geometries were also run with this criterion. Finally, the predictions were compared amongst them and with experimental data from previous works [13, 29].

2 Joint Materials

The mechanical properties of adhesives and substrates obtained from previous works, together with experimentally determined P_{max}, were the starting point for the elastic-plastic analyses of the SLJs. A brief description of how the experimental data was acquired is presented next.

The adhesives considered in this work are the Araldite® 2015 and SikaForce ® 7888. Moreover, four L_O were considered: 12.5 mm, 25.0 mm, 37.5 mm, and 50.0 mm. The substrates were made of aluminium AW6082 T651 plate with a thickness of 3 mm, which has a Young's modulus (E) of 70.07 ± 0.83 GPa, a yield strength (σ_y) of 261.67 ± 7.65 MPa, a $\sigma_u = 324.00 \pm 0.16$ MPa, a failure strain (ϵ_f) of 21.70 ± 4.24, and a Poisson's ratio (ν) of 0.30 [30]. The adhesives were also characterised in previous works [11, 31–33] and their mechanical properties (for tensile and shear) are listed in Table 1.

The substrates were cut to a length that allowed a distance of 180 mm between the Universal Testing Machine (UTM) clamps for all L_O and a width of 25 mm, as described in references [13, 29]. Sets of five specimens were prepared and left to cure for at least 48 h at room temperature. Then, all the joints were experimentally

Table 1 Mechanical properties of the adhesives [11, 31–33]

Property	2015	7888
E (GPa)	1.85 ± 0.21	1.89 ± 0.81
ν	0.35^a	0.35^a
σ_y (MPa)	12.63 ± 0.61	13.20 ± 4.83
σ_f (MPa)	21.63 ± 1.61	28.60 ± 2.0
ϵ_f (%)	4.77 ± 0.15	43.0 ± 0.6
G (GPa)	0.56 ± 0.21	0.71^b
τ_y (MPa)	14.6 ± 1.3	–
τ_f (MPa)	17.90 ± 1.8	20^a
γ_f (%)	43.90 ± 3.4	100^a

[a] Manufacturer's data
[b] Calculated by Hooke's Law

tested using a UTM Shimadzu AG-X 100 with a 100 kN load cell. The tests were performed at 1 mm/min, as described in previous works [13, 29]. The average failure load from each set was considered as the P_{max}.

3 Natural Neighbour Radial Point Interpolation Method

First a general geometry, which represents the problem domain ($\boldsymbol{\Omega}$), is discretised in a set of 'n' nodes $\boldsymbol{N} = \{x_1, x_2, \ldots, x_n\} \in \mathbb{R}^d$. With the NNRPIM, the problem domain Ω can be discretised using either a uniform or randomly distributed nodal set. This method does not require a pre-established mesh; however, it requires to establish the 'influence-cells' ($\boldsymbol{V_i}$), which are created through Voronoï's tessellation from $\boldsymbol{N}$ [22]. Considering a point $\boldsymbol{x_I}$ from $\boldsymbol{\Omega}$ as the point of interest, all the nodes belonging to $\boldsymbol{V_i}$ have an effect on $\boldsymbol{x_I}$. Thus, a variable value at $\boldsymbol{x_I}$ $(u(\boldsymbol{x_I}))$ can be expressed as the sum of one radial based function (RBF) and one polynomial function (P), as shown in Eq. (1) [22, 23, 34].

$$u(\boldsymbol{x_I}) = \sum_{i=1}^{n} R_i(\boldsymbol{x_I})\, a_i(\boldsymbol{x_I}) + \sum_{j=1}^{m} P_j(\boldsymbol{x_I})\, b_j(\boldsymbol{x_I}), \tag{1}$$

where $R_i(\boldsymbol{x_I})$ is the RBF with its non-constant coefficient $a_i(\boldsymbol{x_I})$ [21]. Similarly, $P_j(\boldsymbol{x_I})$ is the polynomial function with its non-constant coefficient $b_j(\boldsymbol{x_I})$ [34]. The NNRPIM uses the multi-quadratics RBF, which is related to the 'Euclidean norm' (r_{Ii}) between $\boldsymbol{x_I}$ and the nodes belonging to $\boldsymbol{V_i}$, as shown in Equation (3).

$$R(r_{Ii}) = \left({r_{Ii}}^2 + c^2\right)^p \tag{2}$$

$$r_{Ii} = \sqrt{(x_i - x_I)^2 + (y_i - y_I)^2 + (z_i - z_I)^2} \tag{3}$$

For parameters c and p values of $c \approx 0.0001$ and $p \approx 0.9999$ were previously determined [21, 23]. The non-constant coefficients $a_i(\boldsymbol{x_I})$ and $b_j(\boldsymbol{x_I})$ can be obtained by applying the interpolation function (1) to all nodes (n) in the domain Ω [23, 34].

Equation (1) can then be represented as a matrix [22]:

$$\boldsymbol{u_s} = \boldsymbol{Ra} + \boldsymbol{Pb}, \tag{4}$$

being

$$\boldsymbol{R} = \begin{bmatrix} r_1(\boldsymbol{x}_1) & r_2(\boldsymbol{x}_1) & \cdots & r_n(\boldsymbol{x}_1) \\ r_1(\boldsymbol{x}_2) & r_2(\boldsymbol{x}_2) & \cdots & r_n(\boldsymbol{x}_2) \\ \vdots & \vdots & \ddots & \vdots \\ r_1(\boldsymbol{x}_n) & r_2(\boldsymbol{x}_n) & \cdots & r_n(\boldsymbol{x}_n) \end{bmatrix} \tag{5}$$

$$\boldsymbol{P} = \begin{bmatrix} p_1(\boldsymbol{x}_1) & p_2(\boldsymbol{x}_1) & \cdots & p_m(\boldsymbol{x}_1) \\ p_1(\boldsymbol{x}_2) & p_2(\boldsymbol{x}_2) & \cdots & p_m(\boldsymbol{x}_2) \\ \vdots & \vdots & \ddots & \vdots \\ p_1(\boldsymbol{x}_n) & p_2(\boldsymbol{x}_n) & \cdots & p_m(\boldsymbol{x}_n) \end{bmatrix} \tag{6}$$

Afterwards, Equation (1) is expressed as:

$$u(\boldsymbol{x}_I) = \left\{\boldsymbol{r}(\boldsymbol{x}_I)^T ; \boldsymbol{p}(\boldsymbol{x}_I)^T\right\} \boldsymbol{M}_T^{-1} \begin{Bmatrix} \boldsymbol{u}_s \\ \boldsymbol{z} \end{Bmatrix} = \left\{\Phi(\boldsymbol{x}_I)^T ; \Psi(\boldsymbol{x}_I)^T\right\} \begin{Bmatrix} \boldsymbol{u}_s \\ \boldsymbol{z} \end{Bmatrix} \tag{7}$$

$$\boldsymbol{M}_T = \begin{bmatrix} \boldsymbol{R} & \boldsymbol{P} \\ \boldsymbol{P}^T & \boldsymbol{Z} \end{bmatrix} \tag{8}$$

$$Z_{ij} = 0 \tag{9}$$

where $\boldsymbol{M}_T$ is the total moment matrix, which includes $\boldsymbol{R}$ and $\boldsymbol{P}$ for solving the problem matricially (Eq. 9) [22], and so Eq. (7) takes the form:

$$u(\boldsymbol{x}_I) = \sum_{i=1}^{n} \Phi_i(\boldsymbol{x}_I)\, u_i, \tag{10}$$

being $\Phi_i(\boldsymbol{x}_I)$ the value of the shape function of interest point $\boldsymbol{x}_I$ at node n_i [22]. Further details of the method are given in References [21–23, 34]. Finally, the material matrix and natural and essential boundary conditions where applied using the weak-Galerkin form [21]; from the equilibrium condition between the stress tensor σ and the body forces b (Eq. 11):

$$\nabla\sigma + b = 0, \tag{11}$$

defining the natural and essential boundary conditions:

$$\begin{cases} \sigma\mathbf{n} = \bar{\mathbf{t}} \\ \mathbf{u} = \bar{\mathbf{u}} \end{cases}, \tag{12}$$

where $\mathbf{n}$ is the normal to the traction boundary, $\bar{\mathbf{t}}$ is the tension applied to the natural boundary (Γ_t), and $\bar{\mathbf{u}}$ is the displacement applied to the essential boundary (Γ_u) [21]. Expressing the weak-Galerkin form in a matricial form (Eq. 13), where $\mathbf{K}$ is the stiffness matrix, the boundary conditions and loading can be applied as:

$$\delta\mathbf{u}\,[\mathbf{K}\mathbf{u} - \mathbf{F}] = 0. \tag{13}$$

3.1 Elasto-plasticity

Once the applied stress is higher than σ_y, the corresponding strain (ϵ) does not follow the Hooke's law. Thus, the strain now have two components, one elastic (ϵ_e) and one plastic (ϵ_p) [14, 21, 35], as in Eq. (14).

$$\epsilon = \epsilon_e + \epsilon_p \tag{14}$$

Although the Hooke's law is not fully fulfilled, there still is a proportionality between stresses and strains, but the increase of the first will be lower in relation to the second [35]. The material stiffness after that point is related to the hardening (κ). In consequence, a yielding criterion (point where yielding starts) is a function of κ [35], as follows:

$$f(\boldsymbol{\sigma}) = k(\kappa), \tag{15}$$

where $f(\boldsymbol{\sigma})$ is a function (mathematical) that relates the yielding to the stress state ($\boldsymbol{\sigma}$), while $k(\kappa)$ is a material property [35]. For metals, the first term of Eq. (15) depends only on the second invariant of the deviatoric stress tensor (J_2'). Thus, $f(\boldsymbol{\sigma})$ takes the form of the von Mises criterion, where $k(\kappa)$ is related to the σ_y, as follows:

$$f(\boldsymbol{\sigma}) = J_2' \tag{16}$$

$$\sqrt{(3)}\sqrt{(J_2')} = \sigma_y. \tag{17}$$

Then, after yielding, the plastic strain (ϵ_p) can be determined using the Prandtl-Reuss flow rule [21, 23, 35], as follows:

$$d\epsilon_p = d\lambda \frac{\partial f(\boldsymbol{\sigma})}{\partial \boldsymbol{\sigma}} \tag{18}$$

$$d\epsilon_p = d\lambda \boldsymbol{a}, \tag{19}$$

where λ is the 'plastic multiplier' and $\boldsymbol{a}$ is the flow vector [21, 35]. Then, the relationship between stress and strain would be:

$$d\sigma = \boldsymbol{c} : \left(d\epsilon - d\epsilon_p\right), \tag{20}$$

where $\boldsymbol{c}$ is the fourth order material constitutive tensor [21, 23], which can be related to Eq. (13). The yield surface for the von Mises yield criterion is then:

$$F(\boldsymbol{\sigma}, \kappa) = f(\boldsymbol{\sigma}) - \sigma_y(\kappa) = 0 \tag{21}$$

$$F(\boldsymbol{\sigma}, \kappa) = \sqrt{(3)}\sqrt{(J_2')} - \sigma_y(\kappa) = 0 \tag{22}$$

Afterwards, the hardening is defined following a linear-elastic and linear-plastic behaviour, also known as bilinear. The hardening is defined from the material properties obtained from traction tests. For the non-linear solution, the procedure described in the literature [21, 28] is followed.

Conversely, yielding of adhesives is function of both J_2' and of the hydrostatic stresses, and so the von Mises criterion is not always suitable. However, this will be addressed later in this manuscript.

3.2 *Exponent Drucker-Prager*

The yield surface governed by the EDP follows an exponent equation (23), in which constants a, b, and p_t are material dependent [8, 15].

$$F = aq^b - p - p_t = 0 \tag{23}$$

$$p_t = a\sigma_y^b + \frac{\sigma_y}{3} \tag{24}$$

$$p = -\frac{1}{3}\left(\sigma_{xx} + \sigma_{yy} + \sigma_{zz}\right) \tag{25}$$

Also, p_t can be expressed as in Eq. (24) when material properties are determined from a uni-axial tensile test; finally, p is the hydrostatic pressure, expressed in Eq. (25) [8, 15]. Substituting (24) into (23) leads to:

$$F = aq^b - p - a\sigma_y^b - \frac{\sigma_y}{3} = 0 \tag{26}$$

For adhesives, an exponent $b = 2$ has been recommended [8, 14], regrouping Eq. (26) and introducing the coefficient λ_{DP}, which relates the material's shear and yield strengths (at the same value of true plastic strain) [8, 14], Eq. (26) takes the form:

$$F = q^2 - 3\sigma_y\left(\lambda_{DP} - 1\right)p - \lambda_{DP}\sigma_y^2 = 0 \tag{27}$$

$$q = \sqrt{3}\sqrt{J_2'}, \tag{28}$$

where q represents the von Mises stress, and so it relates the von Mises criterion with the hydrostatic pressure. It should be noted that, when $\lambda_{DP} = 1$, Eq. (27) is equal to Eq. (21). In addition, Eq. (27) is also equivalent to the formulations described by Dean [8, 14], Raghava [7], and Crisfield [16]. In order to determine λ_{DP}, material testing has to be performed (preferentially, tensile and shear tests). Then, true stresses and strains have to be calculated. Finally, the true plastic strain has to be determined. Ductile and tough adhesives have a quite high failure strain in shear compared to tensile; so it was observed that λ_{DP} has to be determined at points close to the ϵ_f.

Finally, the yield surface for the EDP (Eq. 27) can then be used to calculate the flow vector (***a***) (Eq. 18). Then, the solution procedure is the same as described for the von Mises case. The advantage of having the EDP model as a function of the von Mises stress is that allows to use the von Mises routines together with the EDP ones, and so easing the implementation. Nevertheless, the value of λ_{DP} was calculated for each adhesive and input into the numerical routines for the solution. In addition, if λ_{DP} is not input, the routine solves the problem using the von Mises yield criterion.

3.3 Numerical Models

The geometries used for the experimental testing of the joints were numerically replicated using a custom-written MATLAB® script which distributes nodes, creates elements (for FEM and as background for NNRPIM), assigns material properties, boundary conditions and loads, as described in two previous works [26, 27]. To replicate the experimental set-up, in all the models, the displacement of the left vertical edge was constrained ($U_x = U_y = U_z = 0$), as shown in Fig. 1. Meanwhile, a horizontal displacement (corresponding to the experimental observations) was applied to the right vertical edge, while its vertical displacement was also constrained ($Ux = \delta$, $U_y = U_z = 0$), as shown in Fig. 1. To keep the geometries in correspondence to the experimental work, $t_p = 3$ mm, $t_a = 0.2$ mm, and $L_T = 180$ mm in all cases, and so only L_O changed. In this work, a coarser nodal distribution was used (Table 2) because no effect in the results was found.

P_{max} was determined as the reaction force in the horizontal direction when the maximum principal strain (ϵ_{11}) reaches at its peaks the value of ϵ_f, as proposed

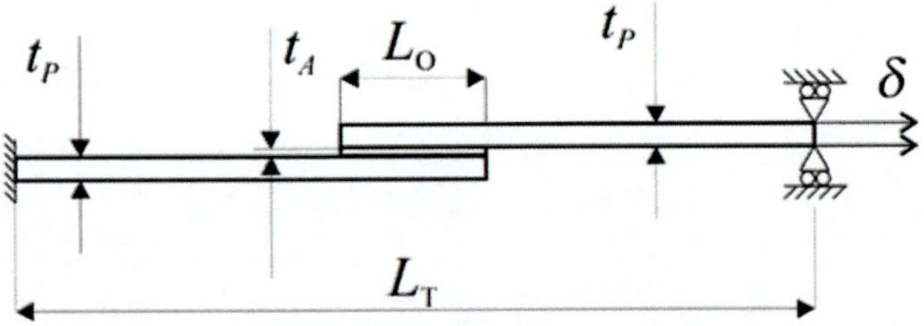

Fig. 1 Geometric definition of the SLJs modelled

Table 2 Substrate and adhesive divisions and biases for nodal distribution

Overlap length	Subs thick	Adh thick	Subs length	Adh length	Length bias (%)	Thick bias (%)	Overlap bias (%)
12.5	10	4	80	102	80	80	20
25.0	10	4	74	188	90	80	40
37.5	10	4	60	220	90	80	40
50.0	10	4	48	254	80	80	40

by Harris and Adams [5, 36] for toughened adhesives, like the SikaForce® 7888. Also, for the Araldite® 2015, the maximum shear strain (γ_{xy}) was utilised instead of ϵ_{11}, as recommended by Hart-Smith [36, 37]; this is due to a better behaviour observed in the Araldite® 2015 using the latter criterion. Also, this adhesive has a rather dissimilar strain strength between tensile and shear in comparison to the SikaForce® 7888 which has similar strengths in both directions.

4 Results

4.1 Experimental

Once all the experimental specimens were tested and their strengths determined, all the substrates were inspected for plastic deformation. No evidence of plastic deformation was observed in the substrates, except those with $L_O = 50$ mm and bonded with the SikaForce® 7888. These joints, in particular, had the highest strength. Moreover, an inspection was done in the adhesive area to determine the type of failure; a cohesive failure was observed in all cases, indicating the correct preparation of the specimens.

In all cases, an increase in P_{max} was observed as L_O increased (Fig. 2). At the shortest L_O, the two adhesives presented rather similar P_{max}, but the SikaForce® 7888 shown the highest. Nevertheless, adhesive ductility has an important effect on longer overlaps because it allows the joint to sustain larger loads.

Fig. 2 Experimental joint strengths from the tested adhesive type-L_O combinations

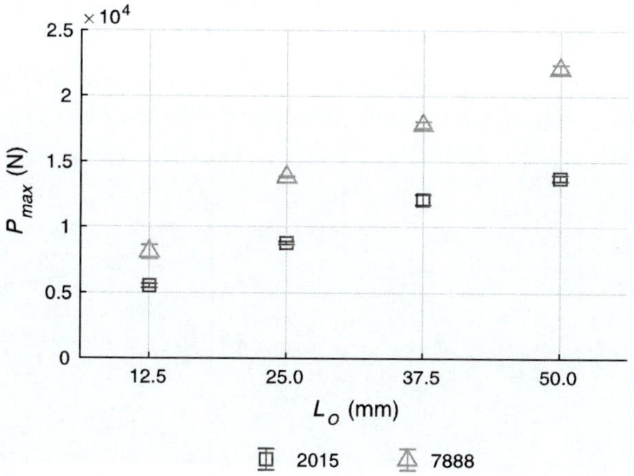

4.2 Stress Distribution

Once the models were solved, the peak of ϵ_{11} was used as a failure criterion when it reached ϵ_f for the SikaForce® 7888. This was performed for both EDP and von Mises criteria and all L_O cases, as shown in Fig. 3. Although the patterns look similar, as well as the peaks, it can be observed that for the smaller L_O, the EDP criterion gives a slightly higher strain along the overlap. Nevertheless, in the other cases, the EDP predicted lower strains in comparison to the von Mises predictions (Fig. 3). For example, the 37.5 mm case, presented an almost zero strain (0.009) at its centre point. Stress distributions corresponding to the SikaForce® 7888 are described because this adhesive is more representative for the method due to its high ductility (Table 1). Nevertheless, the stress distributions corresponding to the Araldite® presented similar trends to those described below.

Fig. 3 Maximum principal strains (ϵ_{11}) of the joints bonded with SikaForce® 7888. Joint strength was considered when $\epsilon_{11} = \epsilon_f$ at its peaks

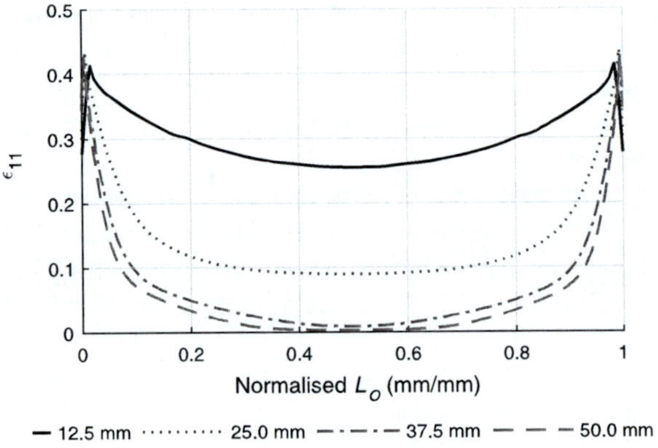

(a) EDP yield criterion

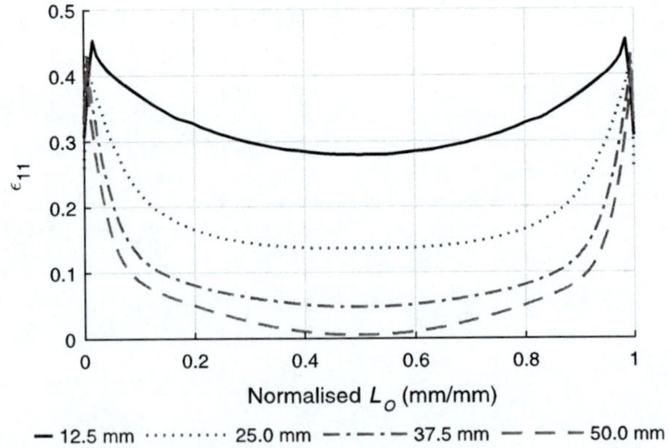

(b) von Mises yield criterion

Fig. 4 Shear stress distribution in the joints bonded with SikaForce® 7888. The distribution from all L_O are superimposed at P_{max}

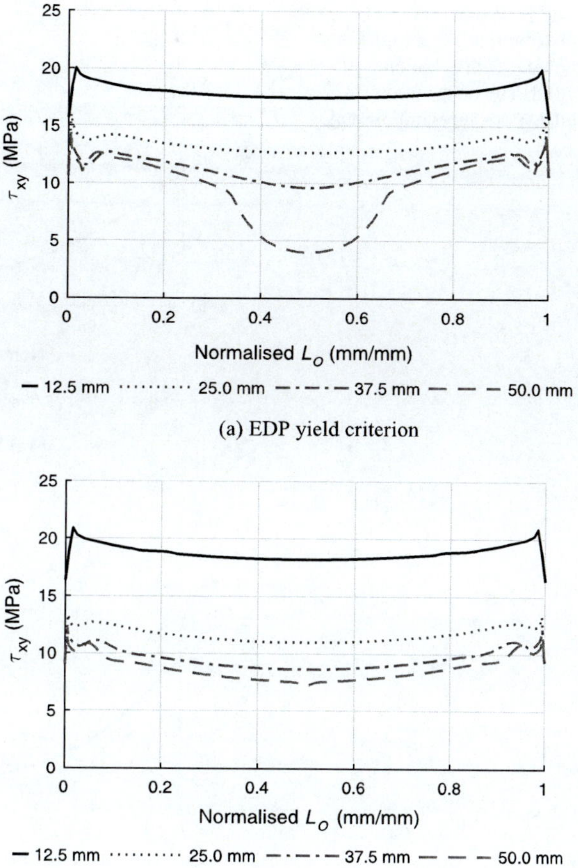

(a) EDP yield criterion

(b) von Mises yield criterion

Shear stresses (τ_{xy}) at the predicted P_{max} presented an approximately constant magnitude along L_O for the 12.5, 25.0, and 37.5 mm cases (Fig. 4a). However, the 50.0 mm case presented a small bump with $\tau_{xy} = 3.99$ MPa indicating that the adhesive layer has not yielded in its total length. Conversely, in the models run using the von Mises criterion, such low spot is not observed (Fig. 4b).

By comparing both Fig. 4a, b, at the centre of L_O, it can be seen that in the smallest L_O there is an slight decrease on τ_{xy} of 17.45 and 18.18 MPa between the EDP and von Mises criteria, respectively. Nevertheless, in the 25.0 mm case, the EDP criterion provided a slight increase on the τ_{xy} to 12.67 from 10.66 MPa. A similar trend was observed in the 37.5 mm case, where τ_{xy} increased to 9.62 from 8.62 MPa. Finally, the 50.0 mm case, presented the lowest stress at its centre point ($\tau_{xy} = 3.99$ MPa). It is important to note that none of the cases surpassed τ_f.

Similarly, the peel stresses σ_{yy} were evaluated, presenting similar trend as τ_{xy} in relation to L_O and yield criterion (Fig. 5). Although no greater difference is observed

Fig. 5 Peel stress distribution in the joints bonded with SikaForce ® 7888. The distribution from all L_O are superimposed at P_{max}

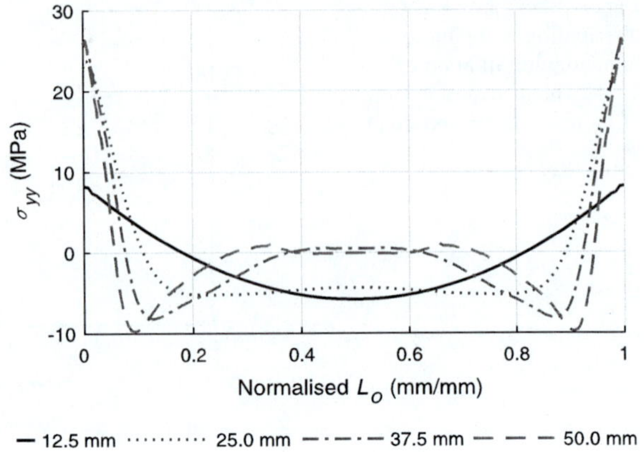

(a) EDP yield criterion

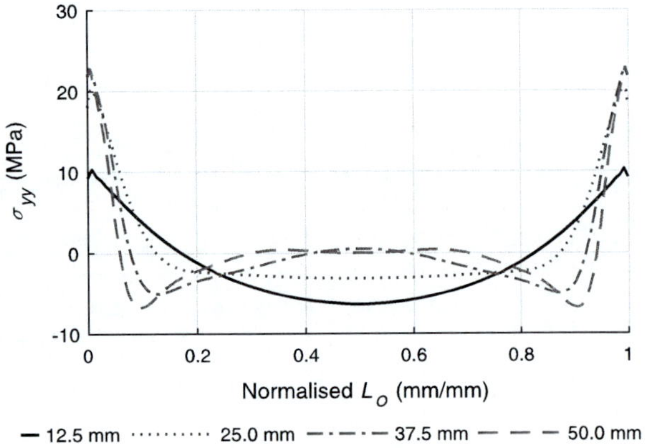

(b) von Mises yield criterion

at the smallest L_O, all the other cases presented higher magnitudes in compression as well as in tension (Fig. 5). Nevertheless, σ_{yy} did not surpass σ_u. Moreover, the two larger L_O presented a region of around $0.3L_O$ at their centres were σ_{yy} was negligible, corresponding also with the regions where the 'effective plastic strain' (ϵ_p) was also close to zero (Fig. 6).

For the same level of joint extension (pulling) the EDP presented lower stresses, in both peel and shear, than the solution using von Mises criterion. Although the lower stresses may disconcert, it seems that the joint could resist more pulling before reaching a 'critical' level (e.g., maximum stress or strain), and so the criterion represents better the adhesive ductility. Similar behaviour was observed in the analyses performed by Dean and Crocker [8], where the EDP predicted stresses were around 30% lower compared to von Mises.

Fig. 6 Effective plastic strain of the joints bonded with SikaForce ® 7888 at P_{max}

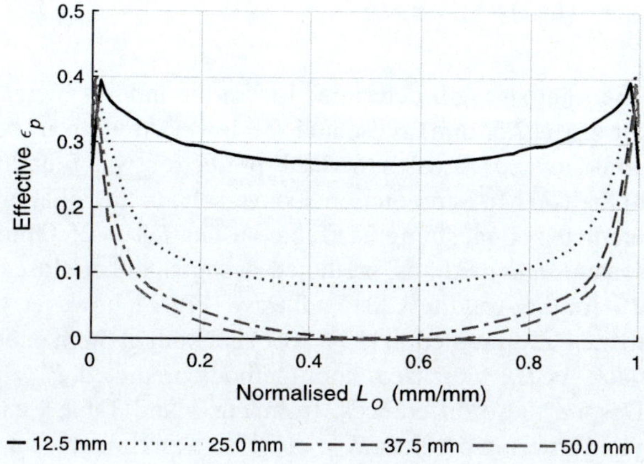

(a) EDP yield criterion

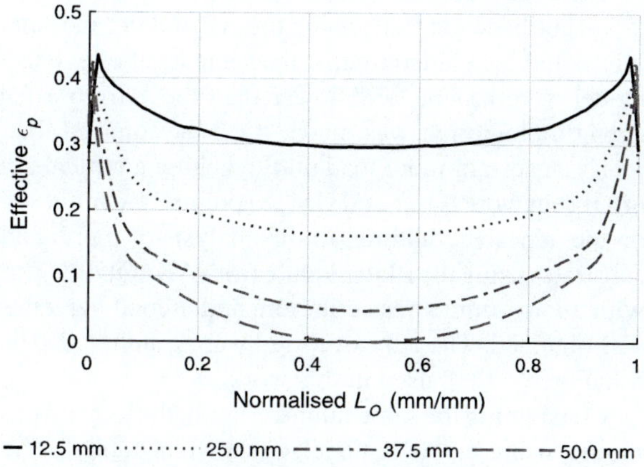

(b) von Mises yield criterion

Brief descriptions of how to include the linear Drucker-Prager and Coulomb-Mohr criteria are available in the plasticity literature [16]. However, a methodology as such to implement the EDP criterion was not found; thus, during its implementation to the meshless method test results were compared with those provided from ABAQUS®, and a good agreement between both methods was found.

In summary, the models solved using the EDP criterion reached higher shear and peel stresses, as shown in Figs. 4 and 5 and larger final displacement (Table 4) in relation to those solved using von Mises.

4.3 Joint Strength

The joint strength determined from the models increased with the L_O. The shortest L_O (12.5 mm) presented the lower strength in both adhesive systems. In the Araldite® 2015s case, the EDP prediction was found to be 10% higher in relation to the von Mises prediction. Conversely, in the SikaForce® 7888s case, both models predicted equal strengths (Table 3). The $L_O = 25.0$ mm models, with both adhesive systems and methods, predicted strengths rather close to the experimental ones, i.e., 2% for the Araldite® 2015 adhesive (Fig. 7); however, the SikaForce® 7888 is closer to the CZM prediction [13] (10%) but both of them under-predicted the experimental data. As L_O increased, both methods predicted P_{max} lower than the experimental. Despite these differences, from Fig. 7 and Table 3 can be observed that the EDP yield criterion predicts P_{max} closer to experimental data mainly for intermediate L_O. Moreover, in the ductile adhesive, Araldite® 2015, the predictions were still close by 17.6%.

The slight increase in P_{max} for the EDP models was expected. Dean and Crocker [8, 14] utilised the EDP, using the ABAQUS® module, to analyse different adhesive joints, and concluded that, at the same level of extension, the stresses from the EDP model were around 33% lower than those from a von Mises model. Although no strength prediction was made, it can be inferred that it must increase because the joint can sustain more load until reaching a critical stress or strain value. Moreover, in a recent work, Özer and Öz [18] pointed the influence that hydrostatic stresses have on the adhesive yielding, in the analysis of mono and bi-adhesive joints with steel substrates using the EDP module from ABAQUS®. The joint strength was determined with a maximum stress criterion, and a good agreement with the experimental data was obtained. The adhesive used was Araldite® 2015, which is less ductile than the SikaForce® 7888 used in this work.

Considering the same failure criteria, the joint displacement also increased in those models using the EDP model, as shown in Table 4. The shortest L_O, however, for the SikaForce® 7888, achieved the same displacement independently of the method, as with P_{max} (Fig. 4), while for the Araldite® 2015 the increase on displacement was 16.3%. For this particular adhesive, the joint displacement at $\tau_{xy} = \tau_f$ did not change with $L_O \geq 25.0$ mm and presented an increase of 15.7% by using the EDP yield

Table 3 Comparison between strengths (in N) predicted by the NNRPIM models using von Mises and EDP in both adhesive systems

L_O	Araldite® 2015		SikaForce® 7888	
(mm)	von Mises	EDP	von Mises	EDP
12.5	7039.5	7799.4	7854.8	7854.8
25.0	8912.4	9390.7	9668.0	10010.3
37.5	9918.6	10385.2	11020.1	11293.0
50.0	10957.8	11465.3	12258.9	12423.3

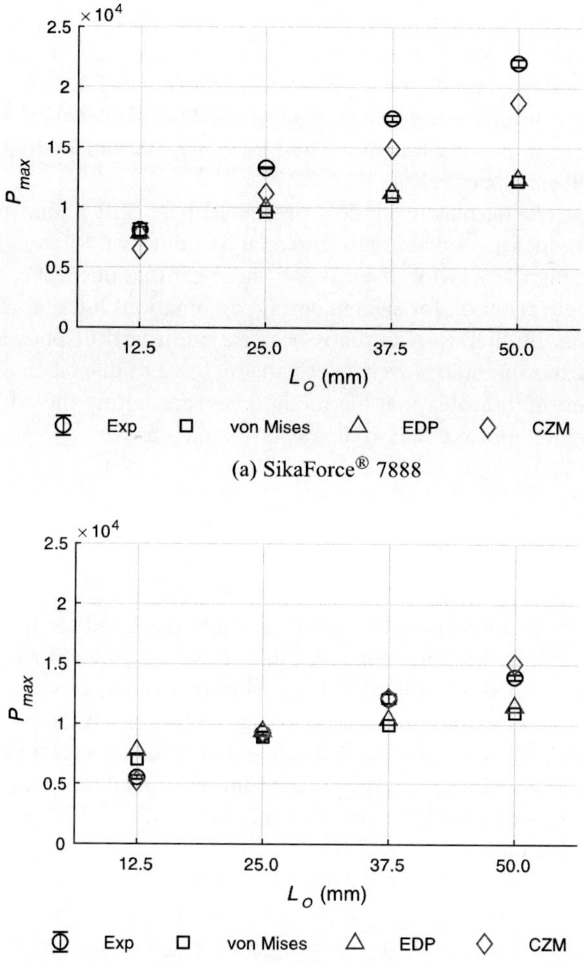

(a) SikaForce® 7888

(b) Araldite® 2015

Fig. 7 Comparison between experimental and CZM [13] with the predictions from the NNRPIM models using von Mises and EDP, respectively

Table 4 Joint displacements (in mm) from the numerical models. Comparison between von Mises and EDP in relation to adhesive system

L_O	Araldite® 2015		SikaForce® 7888	
(mm)	von Mises	EDP	von Mises	EDP
12.5	0.43	0.50	0.61	0.61
25.0	0.51	0.59	0.80	0.95
37.5	0.51	0.59	0.94	1.02
50.0	0.51	0.59	0.99	1.03

criterion in comparison to von Mises (Table 4). In contrast, the models corresponding to the SikaForce® 7888 showed an increase in displacement proportional to L_O; again, the EDP criterion predicted higher values. The $L_O = 25.0$ mm joint presented the higher benefit in this sense because the predicted displacement increased 17% with respect to the von Mises model. The longer L_O presented smaller increments of 7.5% and 4.4%, respectively.

Although the elastic-plastic models presented here still under-predict P_{max} in larger L_O, it provides a good alternative to analyse adhesively-bonded SLJs, mainly, in short and medium L_O. Moreover, these under-estimations indicate the need for defining a strength criterion for each material combination. Joint displacement is not often reported in the literature, perhaps because numerical displacement is always lower than the experimental. Nevertheless, the increase of displacement by using the EDP yield criterion indicates that the method is representing the adhesive (and the joint) slightly more approximate to the experimental cases.

5 Conclusions

The non-linear behaviour of ductile and tough adhesives had led in the direction of non-linear material models for computational analyses, like the FEM. Amongst those non-linear material models is the von Mises criterion that can be successfully applied to some adhesives; however, not to all. During the last two decades, work has been done in the use of other yield criteria, such as Coulomb-Mohr and Drucker-Prager and their variations. Moreover, they are commonly available within commercial FEM software. Evidence exists about the suitability of the EDP yield criterion for adhesively-bonded joints.

Although the FEM is commonly used in the numerical analysis of adhesively-bonded joints, there is still opportunity for other numerical alternatives, such as the meshless methods. Moreover, elastic-plastic analyses are possible using meshless methods. However, the EDP has not been yet implemented. This work aimed to implement the EDP yield criterion using the exponent 'two' into the NNRPIM meshless method, and subsequently to apply it in the analysis of adhesive joints bonded with ductile adhesives. Moreover, the implemented procedure allowed to use the von Mises in the substrates and EDP in the adhesive layer.

The results indicate a promising implementation of the method. First, in longer L_O, the amount plasticised adhesive, at the chosen critical value of strain, was reduced; this effect was not observed with the von Mises criterion. Second, although stresses were higher with the EDP criterion, such increments were more uniform, and so none of them reached failure at the chosen strain level. Third, the predicted joint displacement was higher in relation to the predicted with the von Mises criterion suggesting a better representation of the adhesive ductility. Forth, the criterion and methodology provided P_{max} closer to the CZM with the capability of obtaining stresses and strains within the adhesive layer.

A L_O, in particular, presented the closest behaviour to CZM and experimental data, the 25.0 mm. Also, this joint presented the highest increase in displacement (17%). However, with longer L_O, the method still under-predicts the joint strength suggesting the need for a better failure criterion. Nevertheless, the strengths predicted with this methodology could be used in applications where weight or material savings are not critical because the predictions are in the conservative side. In conclusion, the EDP combined with the NNRPIM meshless method is able to predict joint strength, joint extension, as well as stresses within the adhesive layer and can be used for further analyses of adhesively-bonded joints involving very ductile adhesives.

Funding The authors truly acknowledge the funding provided by the Ministério da Ciência, Tecnologia e Ensino Superior—Fundação para a Ciência e a Tecnologia (Portugal), under project fundings 'MIT-EXPL/ISF/0084/2017', 'POCI-01-0145-FEDER-028351', and 'SFRH/BD/147628/2019'. Additionally, the authors gratefully acknowledge the funding of Project 'NORTE-01-0145-FEDER-000022'—SciTech—Science and Technology for Competitive and Sustainable Industries, co-financed by Programa Operacional Regional do Norte (NORTE2020), through Fundo Europeu de Desenvolvimento Regional (FEDER).

References

1. Adams RD, Wake WD (1984) Structural adhesive joints in engineering. Elsevier Applied Science Publishers LTD, Essex, England. https://doi.org/10.1007/978-94-009-5616-2
2. Fay PA (2005) History of adhesive bonding. In: Adhesive bonding: science, technology and applications. Woodhead Publishing Limited, Cambridge, England, pp 3–22
3. da Silva LFM, das Neves PJC, Adams RD, Wang A, Spelt JK (2009) Analytical models of adhesively bonded joints-Part II: comparative study. Int J Adhes Adhes 29(3):331–341. https://doi.org/10.1016/j.ijadhadh.2008.06.007
4. da Silva LFM, das Neves PJC, Adams RD, Spelt JK (2009) Analytical models of adhesively bonded joints-Part I: literature survey. Int J Adhes Adhes 29(3):319–330. https://doi.org/10.1016/j.ijadhadh.2008.06.005
5. Harris JA, Adams RD (1984) Strength prediction of bonded single lap joints by non-linear finite element methods. Int J Adhes Adhes 4(2):65–78. https://doi.org/10.1016/0143-7496(84)90103-9
6. Bigwood DA, Crocombe AD (1990) Non-linear adhesive bonded joint design analyses. Int J Adhes Adhes 10(1):31–41. https://doi.org/10.1016/0143-7496(90)90025-S
7. Raghava R, Caddell RM, Yeh GSY (1973) The macroscopic yield behaviour of polymers. J Mater Sci 8:225–232
8. Dean GD, Crocker LE, Read B, Wright L (2004) Prediction of deformation and failure of rubber-toughened adhesive joints. Int J Adhes Adhes 24(4):295–306. https://doi.org/10.1016/j.ijadhadh.2003.08.002
9. Mubashar A, Ashcroft IA (2017) Comparison of cohesive zone elements and smoothed particle hydrodynamics for failure prediction of single lap adhesive joints. J Adhes 93(6):444–460. https://doi.org/10.1080/00218464.2015.1081819
10. Fernandes TAB, Campilho RDSG, Banea MD, da Silva LFM (2015) Adhesive selection for single lap bonded joints: experimentation and advanced techniques for strength prediction. J Adhes 91(10–11):841–862. https://doi.org/10.1080/00218464.2014.994703
11. Campilho RDSG, Banea MD, Neto JABP, da Silva LFM (2013) Modelling adhesive joints with cohesive zone models: effect of the cohesive law shape of the adhesive layer. Int J Adhes Adhes 44:48–56. https://doi.org/10.1016/J.IJADHADH.2013.02.006

12. Fernandes RL, Campilho RDSG (2017) Testing different cohesive law shapes to predict damage growth in bonded joints loaded in pure tension. J Adhes 93(1–2):57–76. https://doi.org/10.1080/00218464.2016.1169176
13. de Sousa CCRG, Campilho RDSG, Marques EAS, Costa M, da Silva LFM (2017) Overview of different strength prediction techniques for single-lap bonded joints. Proc Inst Mech Eng Part L J Mater Des Appl 231(1–2):210–223. https://doi.org/10.1177/1464420716675746
14. Dean GD, Crocker LE (2001) The use of finite element methods for design with adhesives. Technical report, National Physical Laboratory, Teddington, Middlesex, UK
15. Simulia User Assistence, ABAQUS 7.0 Documentation (2017)
16. Crisfield MA (1997) Non-linear finite element analysis of solids and structures. Volume 2: Advanced topics. Wiley Ltd, Chichester, West Sussex, UK
17. da Silva LFM, Rodrigues TNSS, Figueiredo MAV, de Moura MFSF, Chousal JAG (2006) Effect of adhesive type and thickness on the lap shear strength. J Adhes 82(11):1091–1115. https://doi.org/10.1080/00218460600948511
18. Özer H, Öz Ö (2017) The use of the exponential Drucker-Prager material model for defining the failure loads of the mono and bi-adhesive joints. Int J Adhes Adhes 76:17–29. https://doi.org/10.1016/J.IJADHADH.2017.02.005
19. Öz Ö, Özer H (2017) An experimental investigation on the failure loads of the mono and bi-adhesive joints. J Adhes Sci Technol 31(19–20):2251–2270. https://doi.org/10.1080/01694243.2016.1264661
20. Nguyen VP, Rabczuk T, Bordas S, Duflot M (2008) Meshless methods: a review and computer implementation aspects. Math Comput Simul 79(3):763–813. https://doi.org/10.1016/J.MATCOM.2008.01.003
21. Dinis L, Natal Jorge RM, Belinha J (2009) The radial natural neighbours interpolators extended to elastoplasticity. In: AJM Ferreira, EJ Kansa, GE Fasshauer, VMA Leitão (eds) Progress meshless methods. Springer Science & Business Media B.V., pp 175–198
22. Belinha J (2015) Meshless methods in biomechanics: bone tissue remodelling analysis. Springer International Publishing. https://doi.org/10.1007/978-3-319-06400-0
23. Farahani BV, Belinha J, Amaral R, Tavares PJ, Moreira PMPG (2019) Extending radial point interpolating meshless methods to the elasto-plastic analysis of aluminium alloys. Eng Anal Bound Elem 100:101–117. https://doi.org/10.1016/J.ENGANABOUND.2018.02.008
24. Tsai CL, Guan YL, Ohanehi DC, Dillard JG, Dillard DA, Batra RC (2014) Analysis of cohesive failure in adhesively bonded joints with the SSPH meshless method. Int J Adhes Adhes 51:67–80. https://doi.org/10.1016/j.ijadhadh.2014.02.009
25. Bodjona K, Lessard L (2015) Nonlinear static analysis of a composite bonded/bolted single-lap joint using the meshfree radial point interpolation method. Compos Struct 134:1024–1035. https://doi.org/10.1016/j.compstruct.2015.08.136
26. Ramalho LDC, Campilho RDSG, Belinha J (2019) Predicting single-lap joint strength using the natural neighbour radial point interpolation method. J Braz Soc Mech Sci Eng 41(9):1–11. https://doi.org/10.1007/s40430-019-1862-0
27. Sánchez-Arce IJ, Ramalho LDC, Campilho RDSG, Belinha J (2019, in Press) Evaluation of an elastic meshless formulation to adhesive joints' strength prediction against established methods. J Adhes Sci Technol, pp 1–27. https://doi.org/10.1080/01694243.2019.1702829
28. Moreira SF, Belinha J, Dinis LMJS, Natal Jorge RM (2017) The anisotropic elasto-plastic analysis using a natural neighbour RPIM version. J Braz Soc Mech Sci Eng 39:1773–1795. https://doi.org/10.1007/s40430-016-0603-x
29. Nunes SLS, Campilho RDSG, da Silva FJG, de Sousa CCRG, Fernandes TAB, Banea MD, da Silva LFM (2016) Comparative failure assessment of single and double lap joints with varying adhesive systems. J Adhes 92(7–9):610–634. https://doi.org/10.1080/00218464.2015.1103227
30. Campilho RDSG, Pinto AMG, Banea MD, Silva RF, da Silva LFM (2011) Strength improvement of adhesively-bonded joints using a reverse-bent geometry. J Adhes Sci Technol 25:2351–2368. https://doi.org/10.1163/016942411X580081

31. Campilho RDSG, Banea MD, Pinto AMG, da Silva LFM, de Jesus AMP (2011) Strength prediction of single- and double-lap joints by standard and extended finite element modelling. Int J Adhes Adhes 31(5):363–372. https://doi.org/10.1016/J.IJADHADH.2010.09.008
32. Campilho RDSG, Moura DC, Gonçalves DJS, da Silva JFMG, Banea MD, da Silva LFM (2013) Fracture toughness determination of adhesive and co-cured joints in natural fibre composites. Compos Part B Eng 50:120–126. https://doi.org/10.1016/J.COMPOSITESB.2013.01.025
33. Neto JABP, Campilho RDSG, da Silva LFM (2012) Parametric study of adhesive joints with composites. Int J Adhes Adhes 37:96–101. https://doi.org/10.1016/J.IJADHADH.2012.01.019
34. Belinha J, Araújo AL, Ferreira AJM, Dinis LMJS, Natal Jorge RM (2016) The analysis of laminated plates using distinct advanced discretization meshless techniques. Compos Struct 143:165–179. https://doi.org/10.1016/j.compstruct.2016.02.021
35. Owen DRJ, Hinton E (1980) Finite Elements in Plasticity, 1st edn. Pineridge Press Limited, Swansea, UK
36. Crocombe AD, Kinloch AJ (1994) Review of adhesive bond failure criteria. Technical reports, AEA Technology INC, Didcot, Oxfordshire, UK
37. Hart-Smith LJ (1973) Adhesive-bonded single-lap joints. Technical reports, NASA, Hampton, Virginia, USA

Techniques for the Mechanical Characterization and Numerical Modelling of Bonded Automotive Structures Under Impact Loads

E. A. S. Marques, P. D. P. Nunes, A. Akhavan-Safar, C. S. P. Borges, R. J. C. Carbas, J. J. M. Machado, M. P. L. Parente, and L. F. M. da Silva

Abstract The automotive industry is currently faced with increasingly strict requirements with regards to crash testing of its structures, which necessitates the development of powerful numerical tools that can reproduce high speed impacts and predict the behaviour of bonded structures under these conditions. The design of adhesive joints for quasi-static loadings is currently well understood, enabling the manufacture of strong, multi-material structures, essential for lightweight and efficient new vehicles. However, the design of impact resistant structures requires the determination of accurate models of the strain rate dependent behaviour of the materials used in vehicle construction. The work recently performed at the Advanced Joining Processes Unit (AJPU) at the Faculty of Engineering of the University of Porto (FEUP) and the Institute of Science and Innovation in Mechanical and Industrial Engineering (INEGI) has focused extensively on processes for high strain rate characterization of the mechanical behaviour of modern crash testing adhesives for the automotive industry. Special testing procedures and experimental equipment have been developed for the determination of strain rate dependent tensile and shear properties as well as fracture mechanics properties such as the fracture toughness. Special attention has been devoted to the study of mixed mode behaviour under impact conditions, with the design of a mixed mode apparatus for impact conditions. To enable the validation of cohesive numerical models based on the experimental data and allow for design optimization procedures, full scale testing of components has also been performed.

E. A. S. Marques (✉) · P. D. P. Nunes · A. Akhavan-Safar · C. S. P. Borges · R. J. C. Carbas · M. P. L. Parente · L. F. M. da Silva
Instituto de Ciência e Inovação em Engenharia Mecânica e Engenharia Industrial (INEGI), Porto, Portugal
e-mail: emarques@fe.up.pt

J. J. M. Machado
Laboratory for Future Interdisciplinary Research of Science and Technology, Tokyo Institute of Technology, Tokyo, Japan

M. P. L. Parente · L. F. M. da Silva
Departamento de Engenharia Mecânica, Faculdade de Engenharia (FEUP), Universidade Do Porto, Porto, Portugal

Lucas F. M. da Silva et al. (eds.), *Industrial Applications of Adhesives*, Lecture Notes in Mechanical Engineering,
https://doi.org/10.1007/978-981-15-6767-4_5

Keywords Impact · Strain rate · Crash resistant adhesives · Mixed mode · Characterization · Adhesives · Adhesive joints

1 Introduction

Considering the ongoing and dramatic climatic changes the world is currently facing, the transportation sector is being mandated to significantly reduce its contribution to CO_2 emissions. This is especially true for the automotive industry, for which many countries and governing bodies have adopted increasingly stricter emission laws, which can only be met by changing the paradigms associated with vehicle construction techniques. Adopting structures with extreme weight reductions is crucial to achieve these goals, but these drastic reductions can only be attained by using carefully engineered multi-material structures (e.g. combining composites plus high strength aluminium alloys) which are quite difficult to assemble using traditional methods of joining such as welding or fastening. Adhesive bonding techniques have thus found a foothold in this sector, as they are relatively inexpensive and, perhaps most importantly, able to join dissimilar materials all while granting adequate strength and stiffness to the vehicle.

Although environmental concerns have been the driving force behind the development of lightweight structures, the automotive industry must still ensure high safety standards for the vehicles it produces. Only by rigorous analysis and testing of adhesive joints and structures it is possible to guarantee that these standards are kept. Most of the existent analysis and testing procedures for structural bonded joints have been established for quasi-static strength testing at room temperature. However, this information is of limited use for automotive manufacturers, which necessitate effective design methods for accurately predicting impact behaviour of structures. However, such models require a thorough understanding of the mechanical and physical processes associated with impact failure, both and the material and joint level.

This work presents a thorough summary on work being carried out to support the design of impact resistant bonded joints for the automotive industry at the AJPU at FEUP/INEGI. The development of advanced testing equipment, especially suited for testing adhesive joints, is first described in detail. The following sections include discussions on the effect of loading rate on the adhesive properties and a describe a series of experimental procedures regarding testing procedures in tensile, shear and fracture toughness in pure mode and mixed mode conditions, all carried out under impact loads. Another section of this chapter also provides a discussion on experimental impact testing techniques at the specimen level, describing procedures and results obtained for a simple joint configuration (the single lap joint—SLJ) and a more complex component level geometry (road vehicle frame header). Lastly, considerations are made regarding work carried out on the numerical modelling of impact behaviour using finite element software associated to cohesive zone modelling. Models for SLJs and component level specimens are discussed, followed by the introduction of strain rate subroutines for modelling adhesive behaviour and a study on the variation of strain rate during fracture toughness testing.

2 Development of Impact Testing Equipment

Several different types of experimental tests can be carried out to assess the impact behaviour of adhesive joints. These tests give important information about the adhesive or the joint tested, allowing to assess the suitability of adhesive or adherend behaviour, quality of surface preparation or joint geometry. One method to categorize these impact tests is by the impact velocity they can provide [1]. Table 1 summarizes the three main categories of impact tests, their corresponding impact velocity and the necessary testing equipment.

The AJPU has carried out development work on two of these machines: a drop-weight impact testing machine and split Hopkinson pressure bar tester (SHPB). The following subsections explain the design work carried out and the relative advantages of the newly developed testing equipment.

2.1 In-House Developed Drop-Weight Testing Machine

The drop weight impact test, performed in a drop weight machine (sometimes also referred to as a drop tower) is performed by simply releasing a drop mass, to which an impactor is attached, from a defined height. The specimen, located below, will be hit by an impactor that carries a kinetic energy equal to the potential energy that the same impactor has before being released [3, 4]. Drop weight machines typically rely only on gravity to provide the desired impact energy, but they can be equipped with an additional system that increases the impact energy by means of a pre-loaded elastic mechanism. A drop weight machine that can apply an impact load to an adhesive joint with a maximum velocity of 5 m/s and a variable mass up to 56 kg was successfully designed, assembled and tested at the AJPU laboratories. This is equipment is shown in Fig. 1a as designed and in Fig. 1b as assembled.

The drop weight machine was designed with various sub-systems that help it operate and acquire data, including an anti-rebound sub-system that uses a novel

Table 1 Different impact tests categorized by velocity [2]

Test classification	Crosshead speed	Suitable test equipment
Low velocity	Up to 5 m/s	Pendulum impact tester
Medium velocity	Between 5 and 10 m/s	Drop weight impact test machine
High velocity	Between 10 and 100 m/s	Split Hopkinson pressure bar tester

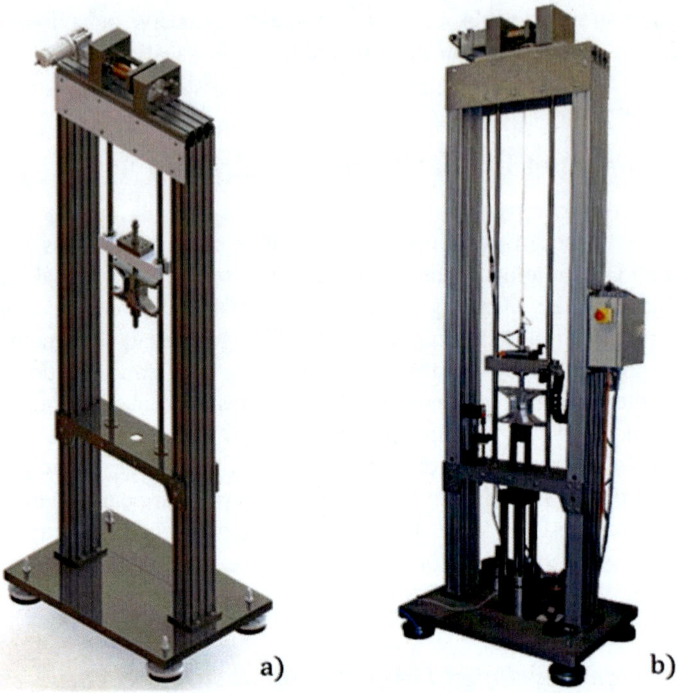

Fig. 1 **a** Drop weight impact testing machine initial design,**b** completed equipment

acting concept and an impactor with a specific shape that adapts to rigid structure (Fig. 2).

The anti-rebound sub-system works by the principle of a lever, ensuring that the anvil and impactor assembly will always be captured, even if the mechanism is not on its final position. The shape of the arms contributes to a quick actuation, taking less time for the arm to start the contact with the anvil.

The U-shaped impactor allows the use of a very rigid specimen fixing structure that avoids deformation and vibration. The obtained data will therefore have minimal contribution of those effects, resulting in a smoother force versus time curve that is only affected by the properties of the tested adhesive joint. A set of piezo electric load cell and accelerometer coupled to the anvil-impactor sub-system provides the necessary experimental data by measuring and recording the force and the acceleration as a function of time. The machine is also equipped with a high frequency laser distance sensor, suitable for detecting small displacements at a rate of up to 80 kHz. This allows for accurate and direct measurement of the displacements that the specimen undergoes, allowing to directly create load-displacement curves that can be compared with those extracted from numerical models.

Fig. 2 Sensors and fixtures installed in the drop weight impact testing machine

2.2 Split Hopkinson Pressure bar Apparatus

As stated in the introduction of this section, testing at velocities between 10 and 100 m/s requires the use of SHPB tester. At these speeds, SHPB testers can generate strain rates on the specimen that range between $10^2\ s^{-1}$ and $10^4\ s^{-1}$ [5]. Although this type of machine was originally developed to test materials under dynamic compression loads, several different variants of the SHPB machine exist, able to carry out tension, torsion and even triaxial and axial/shear combination tests [6]. Given the need for characterisation of adhesives and other materials at strain rates representative of impact conditions, the importance of this type of equipment has only grown. As the AJPU is responsible for various research projects in this field it has embarked in the design of a novel SHPB testing machine.

Although the general layout of a SHPB testing equipment is quite simple, a novel pneumatic actuator system was proposed and designed in this research work. The actuator is designed to ensure that a single SHPB equipment can easily perform compression and tensile impact tests without necessitating significant changes to the machine's geometry. The design of the actuator is shown schematically in Fig. 3, represented in its compression configuration (Fig. 3a) and its tensile configuration (Fig. 3b).

In compression mode, the setup is quite simple as this setup uses the traditional pressure bars, where the striker bar impacts the free end of the incident bar, generating a stress wave that moves towards the specimen, which is located between the incident and the transmission bars. However, in the proposed equipment the striker bar is propelled by the novel pneumatic actuator at high velocities, up to 25 m/s. These extreme velocities necessitate the use of a specially designed braking system for the actuator's rod. This braking system is composed of mechanical springs, that store the rod's kinetic energy until it stops and then restore the elastic energy to the rod,

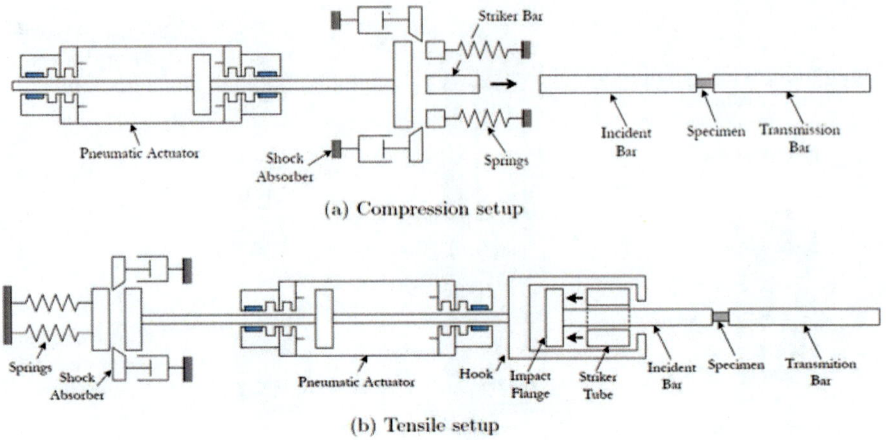

Fig. 3 Schematic representation of the novel pneumatic actuator configured for compression (**a**) and configured for tensile testing (**b**)—Provisonal Portuguese patent PPP 20181000061580 [7]

moving it in the opposite direction until it is stopped by shock absorbers. The tensile setup operates similarly to the compression setup, but the actuator is now commanded to move in the opposite direction, propelling a striker tube attached to the actuator's rod that will then load the specimen.

The design of the actuator required extensive modelling and simulation for validations. In this manner, several mathematical models were used to simulate the various dynamics of the actuation system. Results proved that the proposed system could attain high speeds and is able to launch effectively the striking element that will generate the stress wave. The mechanical design of the novel pneumatic actuator was submitted to provisional patent request [7].

3 Characterization of Adhesives Under High Strain Rates

3.1 Loading Rate and Its Effect on Adhesive Properties

Structural adhesives are mostly polymeric materials, which are known to present a viscoelastic and viscoplastic behaviour. This means that their mechanical properties can vary drastically under different strain rates and temperatures, which has significant implications when designing adhesive joints that will operate under these conditions [8–10]. However, the demand for structures with high strength to weight ratio has led to great developments in adhesive formulations [11], leading to advanced epoxy based adhesives that are able to offer excellent tensile and shear strength under impact conditions. Ashida et al. [12], Nakayama et al. [13] (specifically for the automotive industry). Lutz and Schneider [14] have all proposed novel adhesives which provide increased fracture energy at the expense of a slightly lowered ultimate strength. These are known as crash or impact resistant adhesives and provide unrivalled energy absorption capabilities at the joint level, as their high toughness allows the substrates to deform plastically under impact before the joint fails. To better illustrate the capabilities of these adhesives, Fig. 4 presents a comparison between a brittle and tough adhesive, as well as an example of the extreme plastic deformation that can be attained in a joint with these adhesives.

Accurately characterized mechanical properties for both the adhesives and substrates is thus essential data for the design of joints that will operate under dynamic loadings [2]. Among the most important mechanical properties are the tensile and shear stiffness, the tensile and shear strength, and the fracture toughness (both in mode I, II, or mixed mode). By performing tensile and shear tests at high strain rates, it is possible to derive the basic strain rate dependent properties of the adhesives, such as the stiffness, yield stress and elongation/strain to failure). Ideally, strain rate dependent envelopes for the mechanical properties should be generated, providing data under any impact strain rate and loading mode, as illustrated in Fig. 5.

There is a wide diversity of tests which can be performed using a variety of specific specimens, including tensile tests performed on specimens constituted of bulk material [15], the Thick Adherend Shear Test (TAST) [16, 17], the torsion test [18], or the Notched plate shear method (ARCAN) test [15]. For fracture toughness impact testing, although there are no standardized testing procedures, it is possible to adapt the classical double cantilever beam (DCB) or end notched flexure (ENF) specimens for service under impact loads [19, 20].

The literature presents a large number studies assessing the dynamic behaviour of adhesive joints, but a more limited number of works is devoted solely to adhesives [21–23]. A study of the strain rate effects on the dynamic mechanical behaviour (using a SHPB) of a bi-component epoxy adhesive for structural bonding was performed by Goglio et al. [24], focusing on the influence on the tensile and compressive strength,

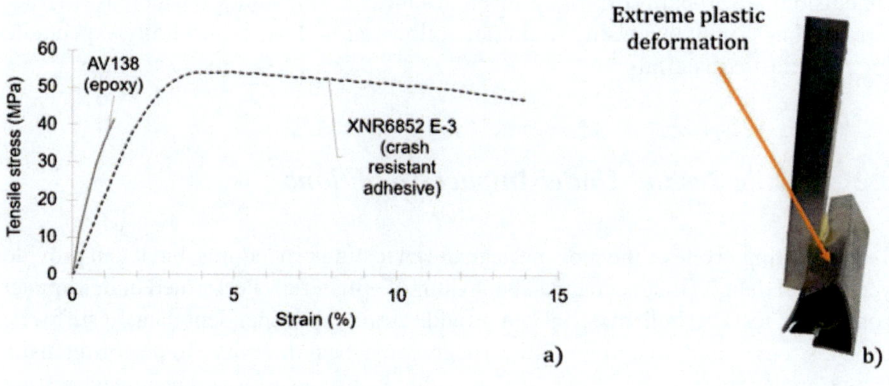

Fig. 4 Comparison of the stress strain behaviour of a stiff, brittle adhesive (AV138) and a modern crash resistant adhesive (XNR6852E-3) (**a**), and visual depiction of the level of plastic deformation possible by the use of a crash resistant adhesive (**b**)

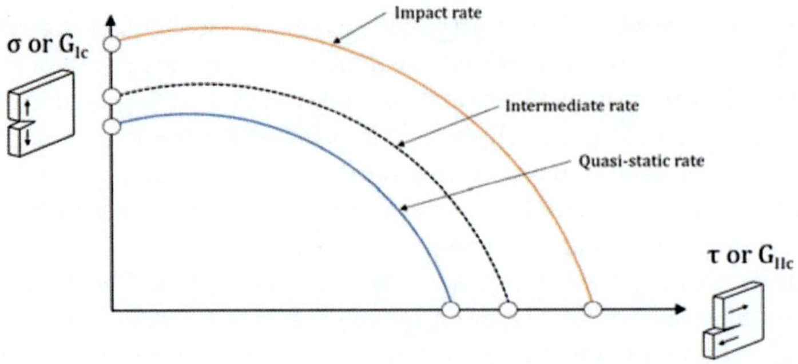

Fig. 5 Schematic representation of a mechanical property envelope as a function of the strain rate

demonstrated that the strength of the adhesive increases significantly with the increase of the strain rate. Goglio et al. [25] performed experimental dynamical testing on two structural epoxy adhesives using a SHPB in tensions and compression, finding a significant effect of the strain rate on the elastic modulus in compression while only a limited effect was found to occur in tension. Using a two-part methacrylate and an epoxy structural adhesives, Yokoyama et al. [26] developed a constitutive model considering high strain rate mechanical compressive properties.

The evolution of the fracture behaviour of adhesives at high strain rates is a more debatable issue, as the results of some authors suggest a decrease in fracture energy in mode I with increasing strain rate while others indicate that it increases. Bascom et al. [27], Bitner et al. [28], Hunston et al. [29], Lataillade et al. [30] and Blackman et al. [31], are among those that have identified a reduction, while authors, such as May et al. [32], Carlberger et al. [33], Marzi et al. [25] and Machado et al. [34] all present experimental works where an increase in fracture energy in mode I is induced by increasing strain rates. Carlberger et al. [33], has even suggested that fracture energy increased linearly in a semi-logarithmic scale. Although these differing conclusions are probably driven by different adhesive behaviours, the increase in fracture toughness with increasing strain rate follows a trend which is wholly expectable for viscoelastic materials.

3.2 Tensile Testing Under Impact Conditions

Tensile testing is one of the most basic material testing procedures, but it can provide very useful data on the mechanical behaviour of a material. Performed under impact conditions, tests on bulk material can provide strain rate dependent data on stiffness, tensile strength and, to a more limited degree, material ductility. To perform tensile testing under impact conditions, it is possible to use a drop-weight impact testing machine or a hydraulic tensile testing machine. The AJPU has performed tests or both cases, but this section will focus on work carried out with an hydraulic tensile

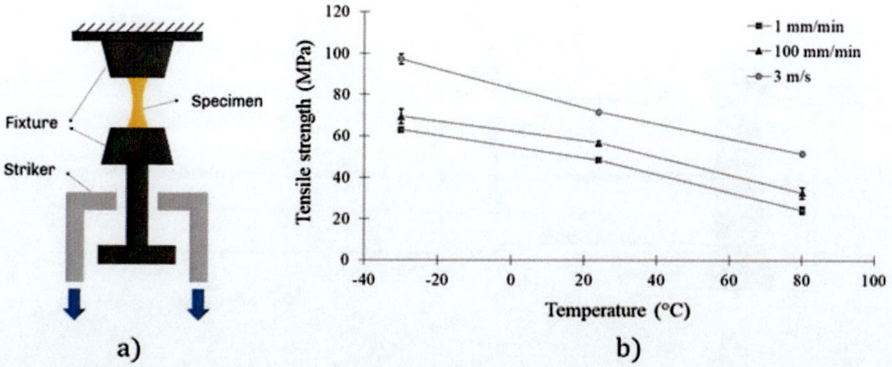

Fig. 6 **a** Schematic representation of a tensile impact test for bulk specimens and **b** representative test results for a crash resistant adhesive (Nagase Chemtex XNR6852E-3)

testing machine, used to test XNR6852 E-3 adhesive bulk at an initial impact velocity of 3 m/s and at −30, 24 and 80 °C (Fig. 12) [34].

To perform this type of tests, it is essential to design a special fixture to hold the specimens. The fixture holds one end of the bulk tensile specimen in place while the lower part of the fixture is attached to the specimen but free to move vertically. A striker is then used to hit the lower section of the fixture at the defined impact velocity. To obtain displacement data, a high-speed camera can be used to record the tests. Tests were performed with this setup at room temperature as well as low (−30 °C) and high temperatures (80 °C). The testing setup is schematically shown in Fig. 6a.

The tensile strength results, shown in Fig. 6b demonstrate that, for this adhesive and for all the displacement rates tested, an increase in temperature results in a decrease of the tensile strength. Similarly, an increase in the strain rate was always found to result in increased tensile strength, independently of the testing temperature. For the adhesive under study, it is thus clear that significant sensitivity exists to both the strain rate and temperature.

3.3 *Shear Testing Under Impact Conditions*

Many researchers argue that tensile testing is not the optimal method for testing an adhesive material, due to the large sensitivity to defects and the non-uniform stress field that is generated at the loaded section [35]. Torsion testing, which obtains shear data is often preferred as a method to assess the true adhesive properties. As described for tensile testing, it is entirely possible to perform torsion tests at high strain rates using, for example, hydraulic torsion testing machines. In this specific machine, a hydraulic torsional actuator is controlled by an open-loop control system due to the low response of closed loop control systems for high strain rates. The adhesive layer

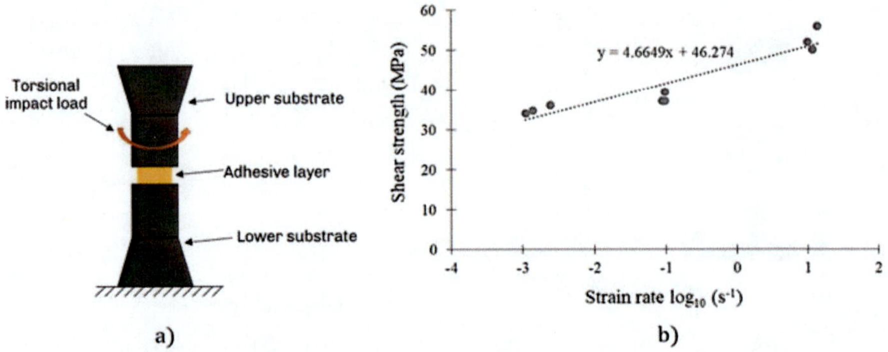

Fig. 7 **a** Schematic representation of a shear impact test for bulk specimens and **b** representative test results for a crash resistant adhesive (Nagase Chemtex XNR6852E-3)

deformation of each specimen was measured using a biaxial extensometer containing two eddy current gap sensors positioned in the axial and circumferential directions. The specimen design consists of two substrates bonded with a circular adhesive layer, as shown in Fig. 7a.

Figure 7b shows an example of the shear strength results obtained using impact torsional testing. To simplify the analysis of the results, the value of strain rate was plotted as $\log_{10}$. The results show relatively low dispersion and clearly show that shear strength increases logarithmically with the value of strain rate.

3.4 Fracture Toughness Testing Under Impact Conditions

Fracture toughness, either in mode I, mode II or mixed mode conditions is perhaps one of the most important adhesive properties for the design of structural bonded joints. In order to measure the critical energy release rate in mode I or mode II, DCB and ENF tests are used, respectively. By testing these specimens, R-curves can be determined after the application of suitable data reduction schemes. Although this type of testing can be carried out at different loading rates, from quasi-static loads to impact with relative simplicity, most conventional data reduction schemes for the determination of fracture toughness rely on the direct measurement of crack location. Although this measurement can be easily performed under quasi-static conditions, crack propagation is extremely fast under impact loads and even high-speed cameras have difficulties in capturing this propagation. Thus, for impact testing it becomes necessary to adopt a data reduction scheme that can be used for this specific case. The compliance-based beam method (CBBM) is a data reduction scheme does not need the exact crack length determination and the method itself considers the fracture process zone. Using this method, fracture energy is given by Eq. 1 [36]:

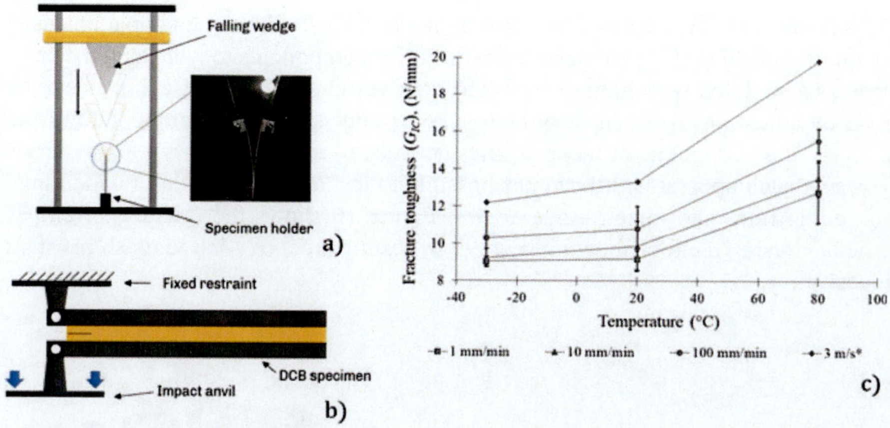

Fig. 8 **a** Schematic representation of a fracture toughness in mode I test using a falling wedge testing setup, **b** schematic representation of a fracture toughness in mode I test using an anvil setup and **c** representative test results for a crash resistant adhesive (Nagase-Chemtex XNR6852E-3)

$$G_I = \frac{6P^2}{B^2 h}\left(\frac{2a_{eq}^2}{h^2 E_f} + \frac{1}{5G_{13}}\right) \tag{1}$$

where P is the load, B is the specimen width, h is the thickness of the substrates, G_{13} is its shear modulus of the adherends, a_{eq} is the length of the crack equivalent and E_f is the corrected flexural modulus, needed because this method does not take into account the stress concentration and the rotation of substrates near the crack tip.

To test DCB specimens under impact, a testing setup similar to those depicted schematically in Fig. 8a, b can be used. In the case of the setup used in Fig. 8a, a falling wedge is used to split the specimen apart during the test. In Fig. 8b the upper substrate of the DCB specimen is pinned to a solid frame while the lower part is connected to an anvil which is free to displace in the vertical direction. When the complete testing jig, including the specimen, is assembled in a drop weight testing machine, an impactor is released and hits the anvil at high speed, generating the necessary impact load to break the specimen.

Figure 8c shows the experimental results obtained by testing the Nagase-Chemtex XNR6852E-3 adhesive under three different levels of temperature and testing speed. The results show that there is a significant increase in the fracture toughness with testing speed, as these crash resistant adhesives significantly increase their energy absorption capabilities with strain rate.

Apart from mode I or mode II testing, data on impact fracture toughness for mixed mode testing is also extremely relevant for industrial applications, namely for the construction of advanced crash test models of complete vehicle structures, which undergo highly complex loading modes at high strain rates.

Mixed mode tests are performed at the AJPU laboratory using an in-house developed apparatus for mixed mode testing under impact loads which uses standard

DCB specimens. This apparatus is based on a previously developed apparatus used for quasi-static tests [37] but designed with stiffer components to withstand the large stresses associated with impact tests. The configuration of the phase angles can be done by adjusting the beam lengths (s_1, s_2, s_3 and s_4). This apparatus allows the analysis of a wide range of mode mixites between 0° and 90° [38].

Using such apparatus, it becomes possible to test adhesives under a wide range of loading rates and mixed mode angles. Figure 10 shows the fracture envelopes generated using the tool shown in Fig. 9b by testing an epoxy-based crash resistant adhesive.

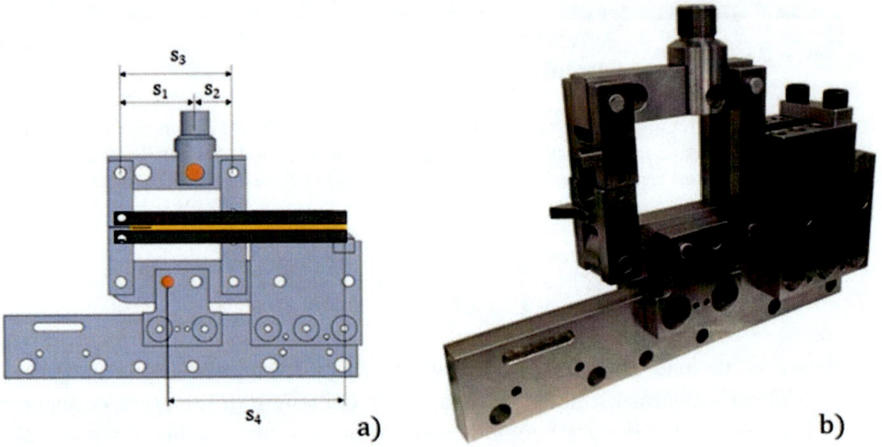

Fig. 9 **a** Schematic representation of the mixed mode, fracture testing apparatus for high strain rates, **b** image of the completed apparatus

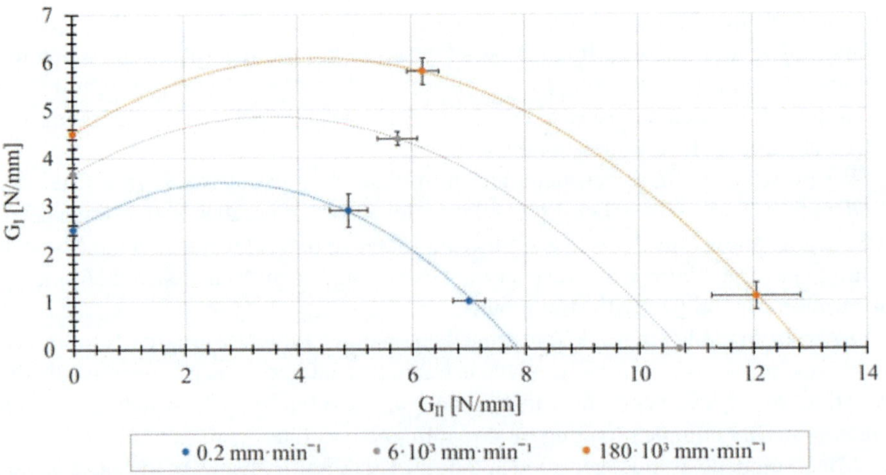

Fig. 10 Fracture envelope for a crash resistant epoxy adhesive under three different loading speeds

An analysis of the fracture envelopes shows that the epoxy-based crash resistant adhesive under study increases its fracture toughness with increasing loading rate for all mode combinations. The data also shows that mixed mode fracture toughness for roughly 45° of loading angle is higher than that found under pure mode I conditions, independently of the loading rate. Data such as this can be directly implemented in material data cards used in large scale crash simulations for the automotive industry, leading to detailed models that are able to reproduce with high fidelity the strain rate dependent behaviour of structural adhesives.

4 Impact Testing of Adhesive Joints

4.1 Specimen Level Testing

Performing specimen level impact testing is a practical method to assess the performance and practical viability of a given adhesive joint configuration. The SLJ is an especially useful joint geometry for this type of test, due to its simplicity and ease of manufacture. Impact tests in SLJs are usually carried out in drop-weight impact testers, with the specimen being mounted vertically with the use of a specially designed and stiff fixture. The impactor of the drop weight tester strikes the lower section of the fixture, accelerating the test assembly so that the specimen is loaded in a tension-shear configuration. The mass and the height of the impactor assembly establish the energy and the speed, respectively, that is applied on the specimen during the impact test.

To illustrate the data that can be gathered from such tests, this section will detail the testing procedure for SLJs with aluminium substrates and composite substrates, bonded with the Nagase-Chemtex XNR6852E-3 crash resistant adhesive. The specimens tested follow the geometrical configuration shown in Fig. 11.

The impact test was carried with a total impactor mass of 26 kg and a test speed of 3 m/s, providing an impact energy of 117 J. This type of tests can be analysed with regards to the failure load and the energy absorbed by the specimen during the testing, which can be approximated by the area underneath the load-displacement curve registered during testing. To visualization, this experimental data can be represented as three-dimensional surfaces. For example, in Fig. 12, two surface plots depict the failure load and energy absorbed as a function of temperature and strain rate for SLJs with composite substrates bonded with Nagase XNR6852E-3 adhesive [39].

The data shows that the while the failure is governed essentially by the strain rate, the energy absorbed is extremely dependent on the temperature and the strain rate, with maximum energy absorption being obtained by high strain rate testing at 80 °C of temperature.

The same analysis was performed for SLJs with aluminium substrates bonded with Nagase XNR6852E-3 adhesive and is shown in Fig. 13.

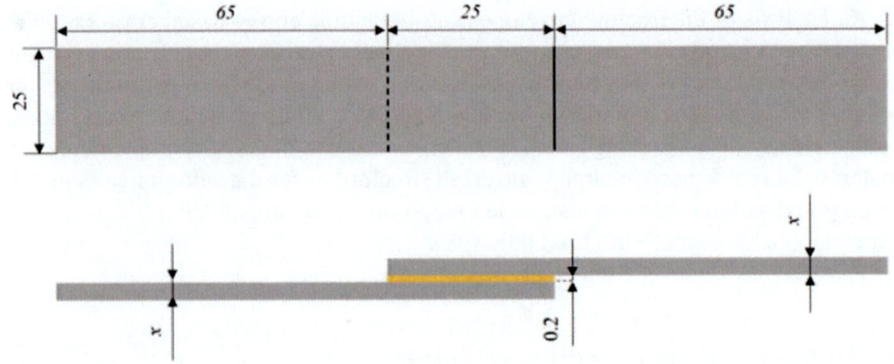

Fig. 11 Schematic representation of a typical single lap joint configuration tested under impact loads (dimensions in mm)

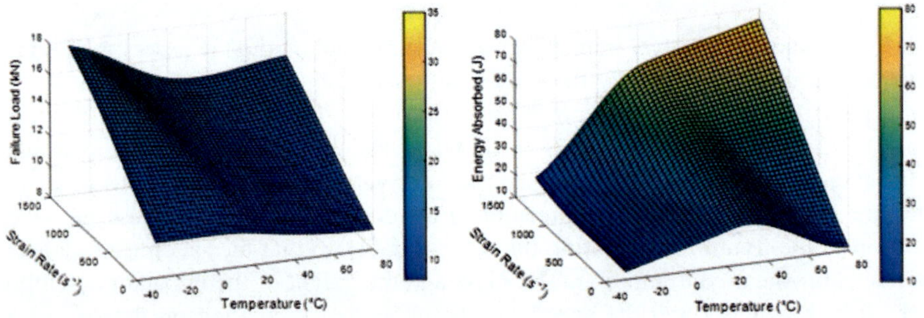

Fig. 12 Failure load and energy absorbed as a function of temperature and strain rate for SLJs with composite substrates bonded with Nagase XNR6852E-3 adhesive

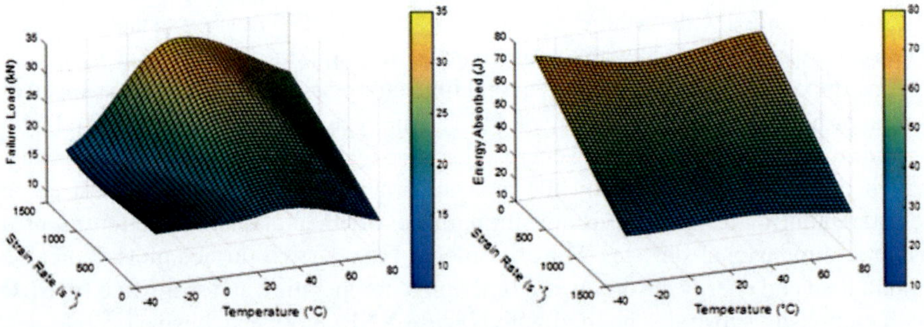

Fig. 13 Failure load and energy absorbed as a function of temperature and strain rate for SLJs with aluminium alloy substrates bonded with Nagase XNR6852E-3 adhesive

In this case, a slight variation of the failure load is found as a function of the temperature, while the energy absorption, while dependent on the strain rate, is almost independent of the temperature, as it is driven by the plastic deformation of the aluminium substrate, which is not temperature dependent at all.

4.2 Component Level Testing

Although small and standardized joint geometries can provide a fast and cost-effective method to study new adhesives and joint configurations, in some cases only testing at the component level, or even full structure testing, can provide the necessary reassurances and validation for more complex cases. However, testing these structures presents interesting challenges, both in the preparation of the experimental setup and the analysis of the resulting data. The AJPU laboratory has extensive experience testing this type of structures, with an example of such a testing procedure being shown in this section. The joint under analysis shown here is a roof frame header from a vehicle. The roof frame header is an aluminium cross frame connection which supports the roof and the windshield of the car (as shown in Fig. 14a) and is composed of an upper and a lower substrate (Fig. 14b), bonded together by two parallel layers of a structural adhesive. Such structural component is of great importance as it should guarantee the passengers safety in the event of a direct impact or in a vehicle rollover. Ideally, the component under impact should deform as far as possible without failing, absorbing most of the impact energy but keeping the integrity of the structure.

In this specific case, Nagase-Chemtex XNR6852 E-3 crash resistant adhesive and a polyurethane adhesive, SikaForce®-7818 L7, were used to bond two substrates, with a nominal layer thickness of 1.6 mm and an average layer width of 17 mm. To create a basis for comparison, quasi-static tests (1 mm/min) were carried out in a servo-hydraulic testing machine. During the test, load is applied to the centre

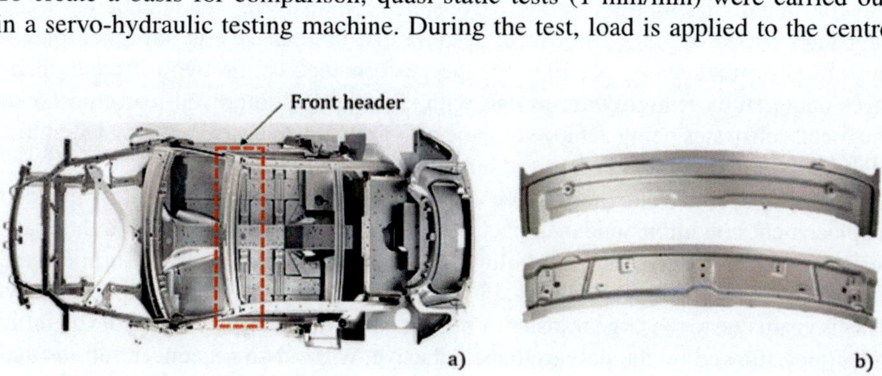

Fig. 14 **a** Front header location on a complete vehicle and **b** the two substrates used in the construction of the front header assembly

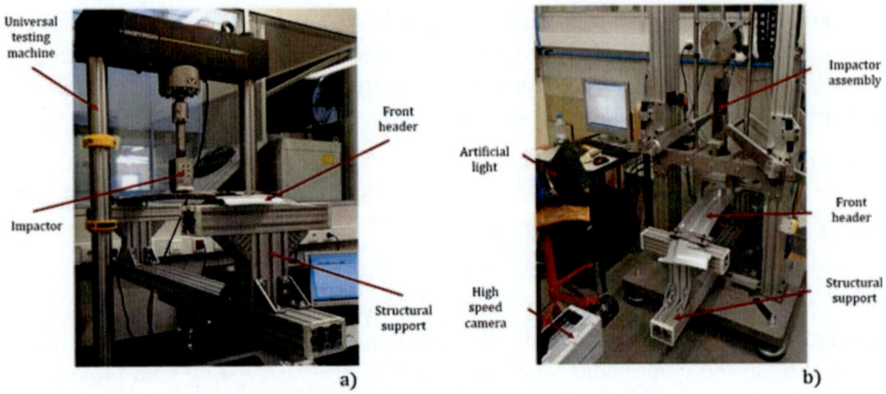

Fig. 15 **a** Quasi-static testing setup and **b** impact testing setup

of the specimens, as shown in Fig. 15a, a configuration generally similar to a 3-point bending test, differing only by not allowing free movement of the specimen outer edges. The acquired data from this test was the load and the displacement. For impact load testing at 3 m/s, the drop weight tester was used with drop mass of 56 kg, released from 1.27 m in height. The full testing setup is shown in Fig. 15b, with the displacement at the centre of the specimen being recorded using a high frequency laser sensor.

During both the impact and the quasi-static tests, the automotive structure specimens were found to fail due to high stress concentration around the holes which are previously drilled in the lower substrate for other vehicle assembly purposes. Such phenomenon led to the initiation and propagation of a crack which caused the decrease of the load carrying capacity of the component. The response of the structure can therefore be said to be mostly dependent of the behaviour of the aluminium substrates, with the adhesive being responsible by transferring the load between the substrates. However, it is an error to assume that the adhesive is not a key player in joint performance, as, per Fig. 16, the performance of the two different adhesives under study is highly dissimilar, with significantly improved load transfer to the metal substrates being achieved by the polyurethane adhesive, achieving a much high higher displacement before sudden failure of the metal occurs.

This is true for both quasi-static and impact conditions. Analysing the load displacement conditions under impact conditions, it is evident that while the crash resistant epoxy can provide high failure load and significant energy absorption, the use of the polyurethane adhesive is able to absorb significantly more impact energy. This is again due to the large transfer of impact energy to the full width of the metallic substrates allowed by the polyurethane adhesive, without stress concentrations that lead to sudden substrate failure (Fig. 17).

This result illustrates the importance of experimental testing as tool to guide the somewhat counterintuitive design choices that often must be made when designing adhesive joint for practical applications. In fact, not only is the joint performance

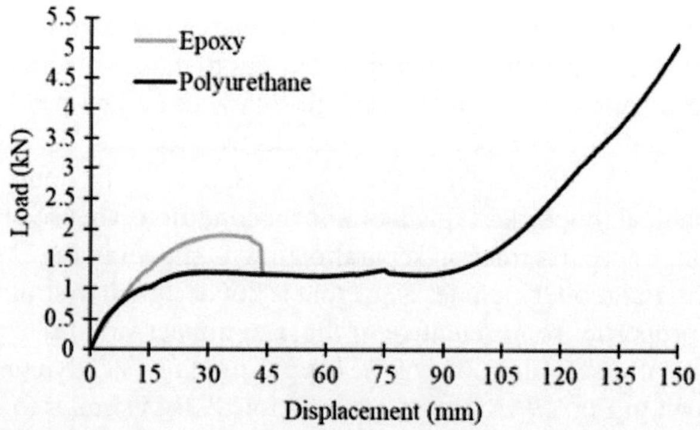

Fig. 16 Load-displacement curve under quasi-static conditions for joints bonded with epoxy and polyurethane adhesives

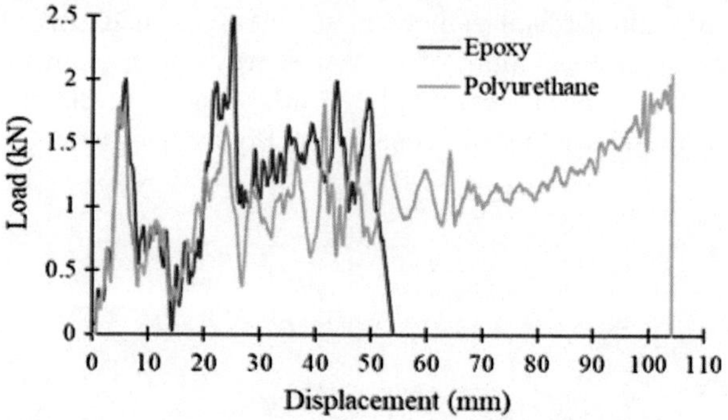

Fig. 17 Load-displacement curve under impact conditions for joints bonded with epoxy and polyurethane adhesives

governed exclusively governed by the adhesive and not only the strongest adhesive can provide the best performance.

5 Modelling the Impact Behaviour of Adhesives and Adhesive Joints

5.1 Cohesive Zone Modelling for SLJs

Cohesive zone models (CZMs) are natively integrated in commercial finite element software packages, such as Abaqus®, providing a powerful tool to simulate damage and failure of an adhesive layer by means of an integrated traction separation law. The most basic of these laws is the triangular traction separation law. This type of law can be modelled with three main parameters for each type of loading mode which are then combined internally by the software according to a user specified mixed mode law [40]. Using a triangular traction separation law, three key properties required for each mode (mode I and mode II) are the elastic modulus, the yield stress and the fracture energy. The triangular CZM can be used the behaviour of SLJs under impact, if mechanical properties typical of impact conditions are used (an example of a high strain rate, triangular traction separation law is shown in Fig. 19. In this case, although the material model includes a law that is not designed to strain rate sensitive it still features properties representative of those at impact strain rate (Fig. 18).

To recreate the impact conditions of the drop weight test 2D dynamic models can be used, as shown in Fig. 19. An impact model for SLJs is similar to those used for quasi-static conditions but features the addition of a mass to the non-clamped edge that represents the impactor mass in the experimental tests.

The boundary conditions follow those typically used in quasi-static models but differ in the application of a velocity field to a mass that is added to the model, instead of a longitudinal displacement in the non-clamped end of the joint. In order to simulate a given amount of impact energy for a given impact velocity, for example 117 J of impact energy with a 3 m/s velocity of the mass, the density in the additional mass is adjusted accordingly. Figure 20 shows a comparison of

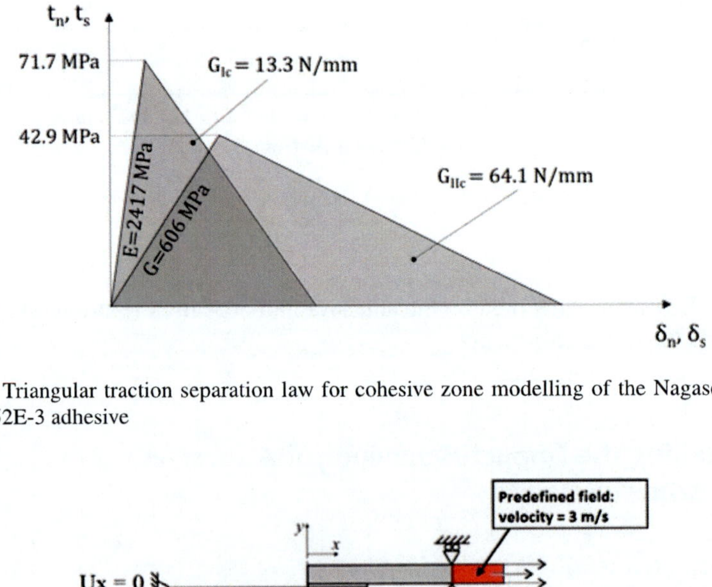

Fig. 18 Triangular traction separation law for cohesive zone modelling of the Nagase Chemtex XNR6852E-3 adhesive

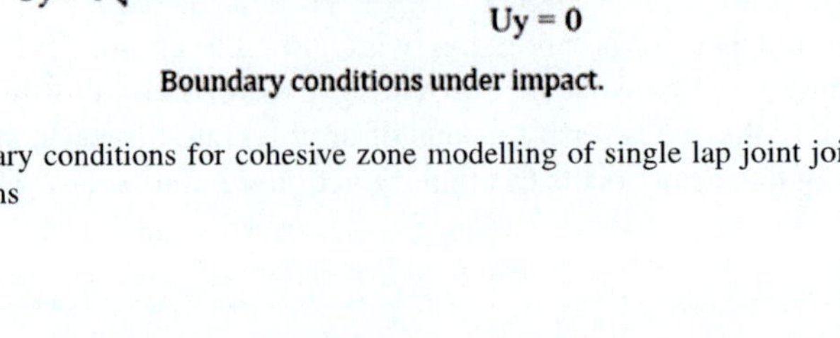

Fig. 19 Boundary conditions for cohesive zone modelling of single lap joint joints tested under impact conditions

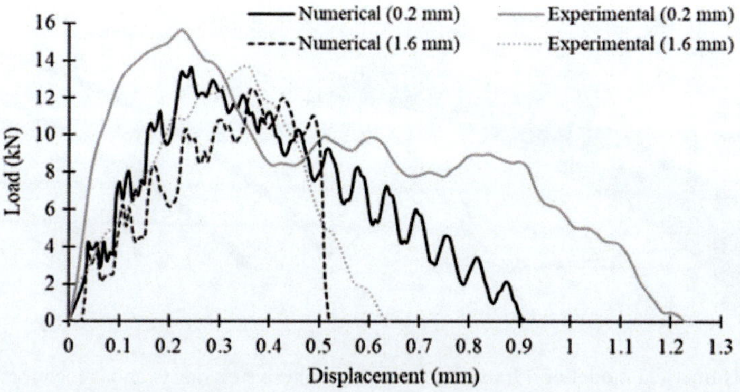

Fig. 20 Comparison of numerical and experimental data for an impact test of aluminium specimens with two different substrate thicknesses and bonded with Nagase-Chemtex XNR6852 E-3 adhesive

numerical and experimental data for an impact test of aluminium specimens with two different substrate thicknesses and bonded with Nagase-Chemtex XNR6852 E-3 crash resistant adhesive.

The results show initial high load peaks induced by large accelerations that occur as contact between the impactor and the anvil is established and do not correspond to the real strength of the joint. The information provided by the load cell must be filtered to remove this undesirable initial noise. After performing this procedure, the failure loads predicted by the numerical models show relatively good agreement with the experimental data although the numerical curves present oscillations which are caused during the simulation of the stress wave propagation.

5.2 *Cohesive Zone Modelling for Bonded Components*

Numerical analysis at the component scale brings additional difficulties to the modelling process. In such cases it is fundamental to ensure that the models for the bonded connections are computationally light, yet sufficiently detailed to capture the failure processes adequately. To provide an example, in this chapter the 3-D model for the roof frame header first presented in Sect. 4.2 is described.

Given the complexity of the model, the model mesh was imported into Abaqus CAE from the Hypermesh® software, where shell elements with thicknesses of 1.7 and 1.2 mm were used, respectively, representing upper and lower aluminium substrates. Most elements were defined as explicit, linear quadrilateral elements (designated by S4 in Abaqus®), although linear triangular elements (S3R in Abaqus®) were used around locations with more complex geometry, as shown in Fig. 21a. Two-parallel adhesive bondlines were modelled each as a single layer of explicit, 8-node three dimensional cohesive elements (COH3D8 in Abaqus®), and shown in Fig. 21b. The mechanical properties for the adhesive were obtained using the exact

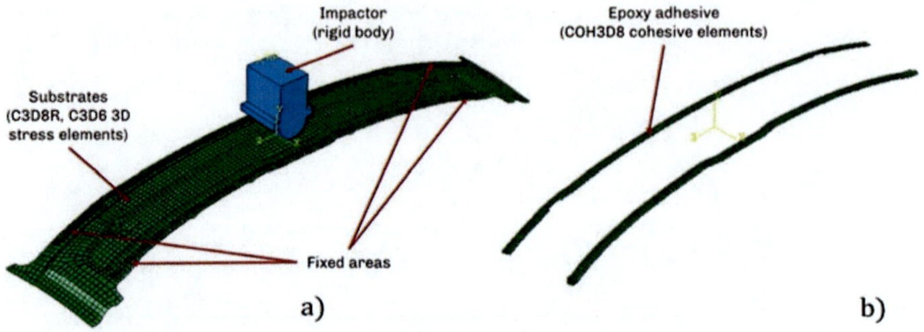

Fig. 21 **a** Numerical model of a frame header using cohesive elements and **b** location of the two adhesive layers modelled by cohesive elements

same procedures described in Sects. 3.2, 3.3 and 3.4 [34]. The boundary conditions of the model included a rigid restraint on both sides of the frame header, roughly simulating how the frame header is connected to the complete vehicle structure. The impactor was given the same mass used in experimental testing and was allowed to move only in the vertical direction, with a maximum displacement of 160 mm. A dynamic, explicit step was used to simulate the impact testing process.

The deformed shape of the structure and the load-displacement curves for the numerical analysis and a representative experimental specimen are shown in Fig. 22.

The numerical model was able to correctly predict the maximum load of the front header, but the displacement was not as accurate. This can be in part attributed to the boundary conditions assigned to the numerical analysis and those verified in the experimental procedure: while in the FE analysis perfect conditions were considered, truly restraining the ends of the component from moving, such condition was not exactly replicated during the tests. In fact, the structure and its assembly during the impact load were less stiff than the numerical model, allowing small movements and therefore leading to higher displacements in the load-displacement curve. In order to compensate these observations, the stiffness of the FE model was slightly corrected to allow a more realistic comparison with the experimental data. Another alternative

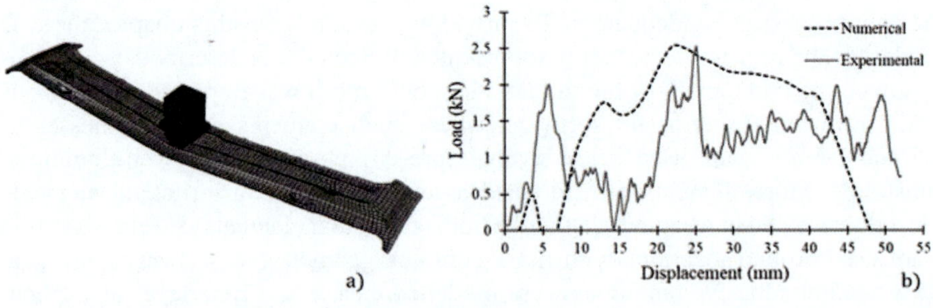

Fig. 22 **a** Deformed shape of the frame header model and **b** comparison of numerical and experimental load-displacement curves (XNR6852E-3 adhesive)

solution could consist in modelling the complete testing setup, including not only the specimen but the structure that holds the specimen during the test. Nonetheless, the performance of the model is still sufficiently accurate to compare energy absorption capabilities which makes it extremely valuable for design purposes.

5.3 Strain Rate Dependent Element Subroutines

As shown before, adhesives exhibit a strain rate dependent behaviour which means that the performance of a bonded joint can only be fully captured by a model that considers the variation of the local strain rate and uses it to update the material properties accordingly. As an example of such procedure, the numerical modelling of DCB tests under impact is shown in this section, with a model that considers strain rate dependent fracture toughness. Achieving this type of strain rate dependent behaviour in a finite element analysis software such as Abaqus, requires the use of a user defined material—UMAT. This type of subroutine allows the determination of the strain rate for each element and the definition of G_{Ic} for each strain rate, through a function previously characterized using experimental data [41].

UMAT defines the Jacobian matrix, $\partial\Delta\sigma/\partial\Delta\varepsilon$, for the mechanical constitutive model, updating stress. A scheme of the integration of UMAT for each increment can be seen in Fig. 23.

Several different cohesive laws can be used for this approach, such as the triangular, trapezoidal and exponential laws [37], however, a triangular law was considered in this approach. The necessary adhesive properties to construct the triangular law are the Young's modulus, E, shear modulus, G_s, the expression that characterizes tensile strength, $\sigma_{\max}$, as a function of strain rate, shear strength, $\tau_{\max}$, the expression that characterizes critical energy release rate in mode I, G_{Ic}, as a function of strain

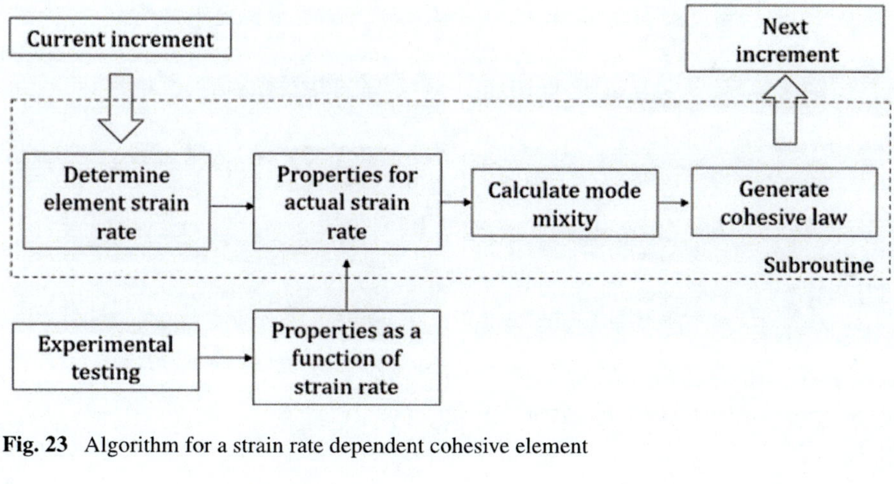

Fig. 23 Algorithm for a strain rate dependent cohesive element

rate and critical energy release rate in mode II, G_{IIc}. The ultimate stress in mode I, $\sigma_{I,\max}$, is the maximum tensile strength $\sigma_{\max}$ and ultimate stress in mode II $\sigma_{II,\max}$ is the maximum shear strength $\tau_{\max}$.

Although this code was developed to analyse cases operating mainly under mode I loads, it is also able to accommodate minor loadings in mode II and the UMAT includes a mixed mode formulation. In this case, for the numerical simulation, the shear modulus G_s was determined from the Young's modulus considering a Poisson's ratio of 0.4 and the G_{IIc} was considered to be five times higher than the G_{Ic} Abaqus® provides, for each time increment, the strain increment and its duration, which allows the determination of strain rate. As mentioned above, if the strain rate is known, the critical energy release rate in mode I can be determined, based on the inputted expression.

The numerical model requires the input of the ultimate tensile stress $\sigma_{\max}$, and energy release rate in mode I, G_{Ic}, as a function of strain rate. As the test progresses, the variation of strain rate can also be seen along the bondline. Therefore, the expressions developed for the properties as a function of strain rate were used. The tests performed were then reproduced using the numerical model.

The numerical model developed has, as output variables, damage, strain rate in mode I and energy release rate in mode I, for each element along the bondline and for each time increment. The results for the model are presented Fig. 24, using a toughened epoxy adhesive loaded at 6×10^3 mm min^{-1}. The crack propagation can be seen, along the test, through damage evolution. When damage is equal to one, $d = 1$, it is considered that the element does not resist the loading anymore, and it is equivalent to the element becoming fully damaged and disappearing.

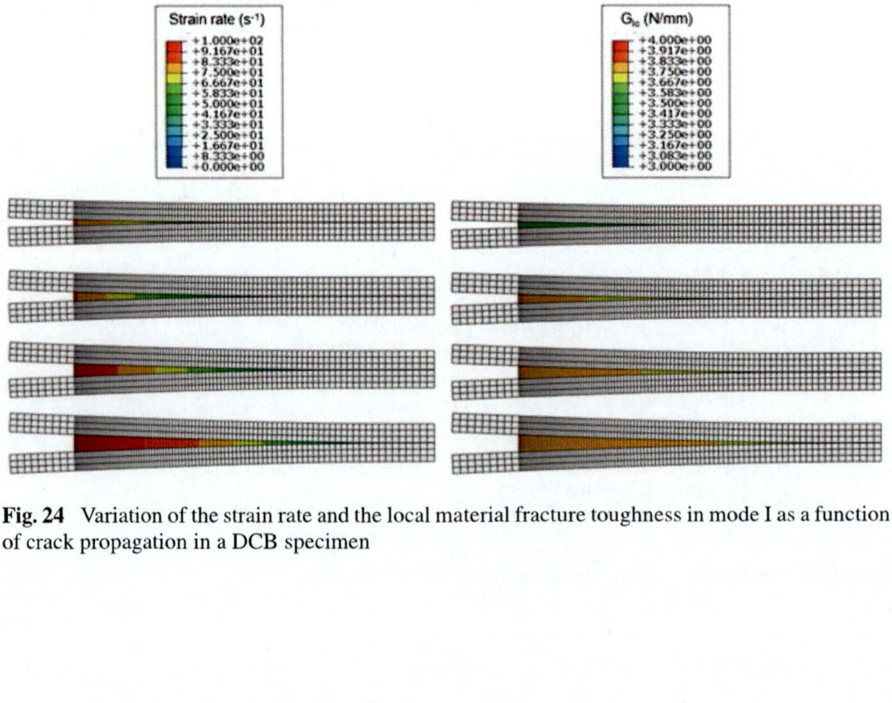

Fig. 24 Variation of the strain rate and the local material fracture toughness in mode I as a function of crack propagation in a DCB specimen

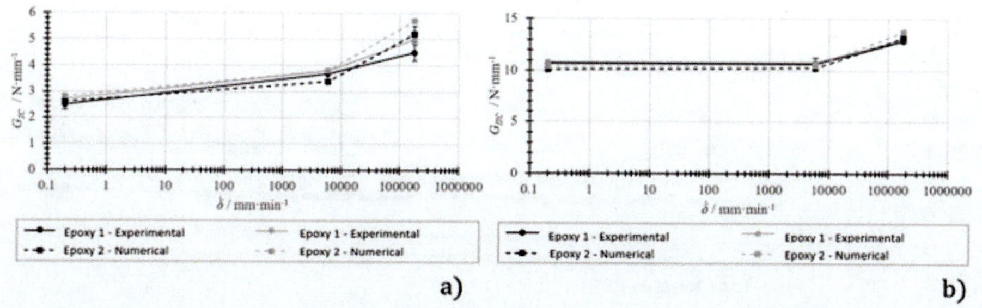

Fig. 25 Comparison of numerical and experimentally obtained fracture values as a function of the strain rate for **a** mode I and **b** mode II loadings

To assess the reliability of the model in the determination of the fracture toughness of DCB and ENF specimens, models were created for two different epoxy-based adhesives and the numerically obtained fracture toughness was compared with experimental data for a wide range of loading rates. These results are shown in Fig. 25.

Overall, the results confirm the validity of the proposed model, as the experimentally and numerically determined failure loads are in general agreement for the complete range of strain rates being tested. It can thus be claimed that such model is suitable for simulating such type of loading and its use is encouraged.

5.4 *Modelling of the Strain Rate Variation in Testing Specimens*

As stated before, strain rate is highly dependent on loading conditions and the specimen conditions. In the case of the propagation of the crack along the specimens of DCB and ENF tests, careful definition of the strain rate is necessary. The general definition of strain rate is expressed in Eq. 2:

$$\dot{\varepsilon}(t) = \frac{d\varepsilon}{dt} = \frac{d}{dt}\left(\frac{L(t) - L_0}{dt}\right) = \frac{v(t)}{L_0} \tag{2}$$

Since strain rate can be defined as velocity at which the ends are moving away from each other, divided by the initial length, it is possible to assess the strain rate of the adhesive layer in mode I for DCB and mode II for ENF using data extracted from a numerical model. This can be achieved by monitoring the displacement as a function of the time increment for the nodes located in the upper and lower limits of each element of the actual adhesive layer. This data can then be plotted considering the engineering and the true strain rate. For a DCB test, the results are shown in Fig. 26a, while the relevant boundary conditions are schematically shown in Fig. 26b

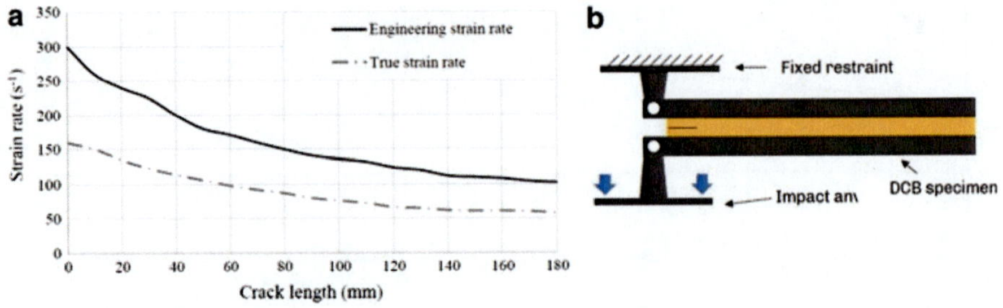

Fig. 26 **a** Variation of the local strain at the crack tip during a DCB test under impact conditions and **b** schematic representation of the testing configuration

Figure 26a indicates that when the adhesive failure occurs, the strain rate varies significantly along the DCB specimens. Both engineering and true strain rate are maximum at the beginning of the test and strain rate starts to decrease in a rapid manner. In addition, throughout the test, the rate at which the strain rate varies decreases. At the end of the test, the value of the strain rate on the adhesive layer is almost one third of the initial value of strain rate. This is very significant variation, especially when the strain rate dependent behaviour of most adhesives is considered.

This data can then be plotted considering the engineering and the true strain rate. For a DCB test, the results are shown in Fig. 27a, while the relevant boundary conditions are schematically shown in Fig. 27b.

Figure 27a plot shows that the strain rate is almost constant at the beginning of test and starts to drop for as the crack length reaches 30 mm. Despite the significant reduction of strain rate at the later stages of testing, this was found to not affect the measurement of the G_{IIc} as stable crack propagation only occurs during this initial 30 mm of displacement. In practice, it can be considered that the ENF specimen cannot be considered as especially sensitive to variations in the strain rate during the test.

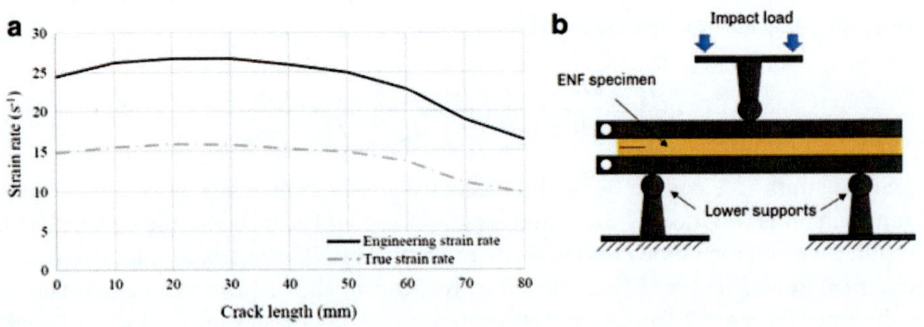

Fig. 27 **a** Variation of the local strain at the crack tip during a ENF test under impact conditions and **b** schematic representation of the testing configuration

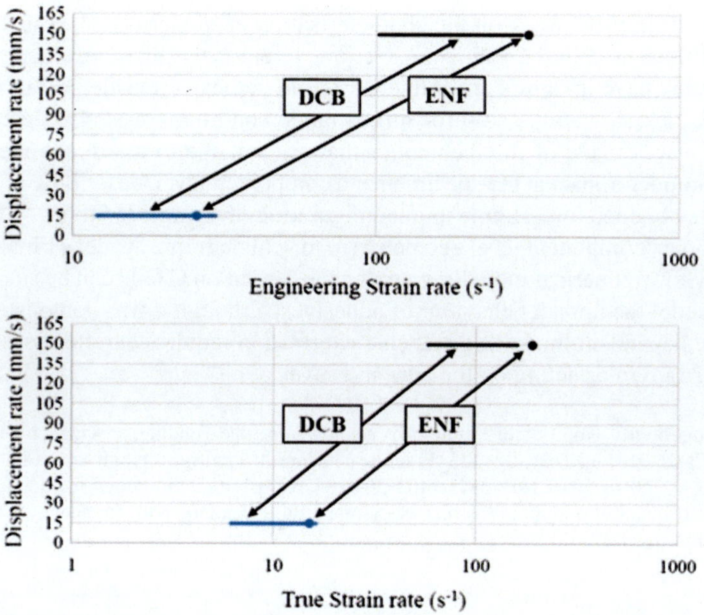

Fig. 28 Plots showing the engineering and true strain rate ranges as a function of the crosshead displacement rate for DCB and ENF tests

Figure 28 shows a compilation of the results of the strain rate as function of the crosshead displacement rate, for both DCB and ENF models, using a crash resistant epoxy adhesive, loaded at 15 and 150 mm/s.

The data indicates that for a given loading rate in ENF tests, the strain rate in the adhesive layer corresponds to the higher end of the range of the strain rate values measured in the DCB models. This work has also shown that if DCB specimens are tested at a fixed crosshead speed, this will load the adhesive layer under a variable strain rate as the test proceeds, which can have important implications in the measured fracture toughness values, while the variation in the strain rate during the ENF test seems to be negligible.

6 Conclusions

This work has presented a detailed summary on work being carried out at the Advanced Joining Processes Unit at FEUP/INEGI to support the design of impact resistant bonded joints for automotive applications. The development of a novel drop weight impact tester and SHPB tester was described in detail, highlighting the special characteristics which are necessary for testing adhesive joints under impact conditions. Extensive work on the characterization of modern crash resistant adhesives was presented in the following section, where it was demonstrated that high loading rates have a significant influence on the adhesive tensile, shear and fracture properties. A testing setup for mixed mode conditions was also described. A description was made of the techniques that are suitable to assess the mechanical behaviour under impact at the specimen and component level, describing procedures and results obtained for a simple joint configuration (the single lap joint—SLJ) and a more complex component level geometry (road vehicle frame header). Finally, it was described how numerical modelling approaches based on CZMs can be successfully used to model the impact behaviour of adhesives, although it was demonstrated that strain rate dependent modelling and characterization requires careful assessment of the local strain rates acting on the adhesive layer.

Acknowledgements The authors gratefully acknowledge the funding provided by Project Nº 29839 POCI-01-0145-FEDER-028473 "Fracture envelope of adhesives at high strain rates (FEASR) and Project Nº 028473 POCI-01-0145-FEDER-028473 Design methodology for impact resistant bonded multi-material automotive structures (ImpactBondDesign), both funded by NORTE-01-0145-FEDER-000022.

References

1. Adams RD (2005) Adhesive bonding: science, technology and applications. Woodhead Publishing, Cambridge
2. Machado JJM, Marques EAS, da Silva LFM (2018) Adhesives and adhesive joints under impact loadings: an overview. J Adhes 94(6):421–452
3. Money MW (1988) Instrumental falling weight impact testing of polymethacrylate and high density polyethylene. Doctoral dissertation. Queen Mary University of London
4. Ambur D, Prasad C, Waters W (1995) A dropped-weight apparatus for low-speed impact testing of composite structures. Exp Mech 35(1):77–82
5. Sharma S, Chavan V, Agrawal R, Patel R, Kapoor R, Chakravartty J (2011) Split-Hopkinson pressure bar: an experimental technique for high strain rate tests. Report. Bhabha Atomic Research Centre
6. Chen WW, Song B (2010) Split Hopkinson (Kolsky) bar: design, testing and applications. Springer Science & Business Media, New York
7. Tenreiro AFGdS CMM, Lopes AMFM, da Silva LFM (2018) Split Hopkinson pressure bar setup for tensile and compression testing with pneumatic impact actuator, Provisonal Portuguese patent PPP 20181000061580
8. Wegman RF, Tanner WC (1964) Strength of epoxy adhesives when stressed to failure in milliseconds. ASTM Spec Tech Publ-Adhes 360(360):163–165
9. Hartshorn SR (1986) Structured adhesives: chemistry and technology. Plenum Press, New York
10. Wake WC (1978) Theories of adhesion and uses of adhesives: a review. Polymer 19(3):291–308
11. da Silva LFM, Öchsner A, Adams RD (eds) (2018) Handbook of adhesion technology, 2nd edn. Springer Science & Business Media, Berlin
12. Ashida T, Ohnishi M, Nagase T, Nakayama A (1994) Epoxy resin adhesive composition. US Patent 5,290,857, 1994
13. Nakayama A, Nagase T, Ashida T, Ohnishi M (1997) Epoxy resin structural adhesive composition. US Patent 5,686,509, 1997

14. Lutz A, Schneider D (2013) Toughened epoxy adhesive composition. US Patent 8,404,787, 2013
15. da Silva LFM, Dillard DA, Blackman B, Adams RD (eds) (2012) Testing adhesive joints: best practices. Wiley, Hoboken
16. da Silva LFM, Adams RD (2005) Measurement of the mechanical properties of structural adhesives in tension and shear over a wide range of temperatures. J Adhes Sci Technol 19(2):109–141
17. Banea MD, da Silva LFM, Campilho RDSG (2012) Moulds design for adhesive bulk and joint specimens manufacturing. J Assem Autom
18. Sato C, Ikegami K (1999) Strength of adhesively-bonded butt joints of tubes subjected to combined high-rate loads. J Adhes 70(1–2):57–73
19. Hayashida S, Sugaya T, Kuramoto S, Sato C, Mihara A, Onuma T (2015) Impact strength of joints bonded with high-strength pressure-sensitive adhesive. Int J Adhes Adhes 56:61–72
20. Machado JJM, Hayashi A, Sekigushi Y, Campilho RDSG, Marques EAS, Sato C, da Silva LFM (2019) Dynamic behaviour in mode I fracture toughness of CFRP as a function of temperature. Theoret Appl Fract Mech 103:102257
21. Harris JA, Adams RD (1985) An assessment of the impact performance of bonded joints for use in high energy absorbing structures. Proc Inst Mech Eng Part C: J Mech Eng Sci 199(2):121–131
22. Yokoyama T (2003) Experimental determination of impact tensile properties of adhesive butt joints with the split Hopkinson bar. J Strain Anal Eng Des 38(3):233–245
23. Martinez MA, Chocron IS, Rodriguez J, Gálvez VS, Sastre LA (1998) Confined compression of elastic adhesives at high rates of strain. Int J Adhes Adhes 18(6):375–383
24. Goglio L, Peroni L, Peroni M, Rossetto M (2008) High strain-rate compression and tension behaviour of an epoxy bi-component adhesive. Int J Adhes Adhes 28(7):329–339
25. Goglio L, Peroni M, Rossetto M (2007) Effect of the strain rate on the mechanical behaviour of epoxy adhesives. In: Key engineering materials. Trans Tech Publ, pp 671–676
26. Yokoyama T, Nakai K, Mohd Yatim NH (2012) High strain-rate compressive properties and constitutive modeling of bulk structural adhesives. J Adhes 88(4–6):471–486
27. Bascom WD, Ting RY, Moulton RJ, Riew CK, Siebert AR (1981) The fracture of an epoxy polymer containing elastomeric modifiers. J Mater Sci 16(10):2657–2664
28. Bitner JL, Rushford JL, Rose WS, Hunston DL, Riew CK (1981) Viscoelastic fracture of structural adhesives. J Adhes 13(1):3–28
29. Hunston DL, Bullman GW (1985) Viscoelastic fracture behaviour for different rubber-modified epoxy adhesive formulations. Int J Adhes Adhes 5(2):69–74
30. Lataillade JL, Grapotte D, Cayssials F (1994) The impact resistance of CTBN-modified epoxy adhesive joints. J Phys IV 4 (C8):C8-771–C8-776
31. Blackman BRK, Kinloch AJ, Sanchez FSR, Teo WS, Williams JG (2009) The fracture behaviour of structural adhesives under high rates of testing. Eng Fract Mech 76(18):2868–2889
32. May M, Voß H, Hiermaier S (2014) Predictive modeling of damage and failure in adhesively bonded metallic joints using cohesive interface elements. Int J Adhes Adhes 49:7–17
33. Carlberger T, Biel A, Stigh U (2009) Influence of temperature and strain rate on cohesive properties of a structural epoxy adhesive. Int J Fract 155(2):155–166
34. Machado JJM, Hayashi A, Nunes PDP, Marques EAS, Carbas RJC, Sato C, da Silva LFM (2019) Strain rate dependence of a crash resistant adhesive as a function of temperature for the automotive industry. Proc Inst Mech Eng Part L: J Mater: Des Appl 233(11):2189–2203
35. Adams RD, Comyn J, Wake WC (1997) Structural adhesive joints in engineering. Springer Science & Business Media, New York
36. De Moura MFSF, Campilho RDSG, Gonçalves JPM (2008) Crack equivalent concept applied to the fracture characterization of bonded joints under pure mode I loading. Compos Sci Technol 68(10–11):2224–2230
37. Costa M, Carbas RJC, Marques EAS, Viana G, da Silva LFM (2017) An apparatus for mixed-mode fracture characterization of adhesive joints. Theoret Appl Fract Mech 91:94–102
38. Borges CSP, Nunes PDP, Akhavan A, Marques EAS, Carbas RJC, Alfonso L, Silva L (2020) Influence of mode mixity and loading rate on the fracture behaviour of crash resistant adhesives. Theoret Appl Fract Mech 107:102508

39. Machado JJM, Nunes PDP, Marques EAS, da Silva LFM (2018) Adhesive joints using aluminium and CFRP substrates tested at low and high temperatures under quasi-static and impact conditions for the automotive industry. Compos Part B: Eng 158:102–116
40. Campilho RDSG, Banea MD, Neto JABP, da Silva LFM (2013) Modelling adhesive joints with cohesive zone models: effect of the cohesive law shape of the adhesive layer. Int J Adhes Adhes 44:48–56
41. Borges CSP, Nunes PDP, Akhavan-Safar A, Marques EA, Carbas RJC, Alfonso L, Silva LFM (2020) A strain rate dependent cohesive zone element for mode I modeling of the fracture behavior of adhesives. Proc Inst Mech Eng Part L: J Mater: Des Appl. https://doi.org/10.1177/1464420720904026

Analysis of Temperature Effect on Deformation Behaviour and Bond Strength of Adhesive Joints with Steel and Composite Substrates

B. Nečasová, P. Liška, and J. Šlanhof

Abstract The supply of materials on the construction market is very broad, almost inexhaustible, in an effort to satisfy the financial and social requirements of all investors. Steel, aluminium and composites are currently the most commonly used materials in construction applications. The joining of various parts and components is usually ensured by traditional mechanical or welded joints, the efficiency and durability of which has been already proven many times. On the other hand, adhesive joints proved their efficacy too and moreover, they allow the implementation of flexible joints that provide a solution with uniform stress distribution. This article presents the results of the stress/strain behaviour of flexible adhesive joints with steel and composite substrates that are very often brittle; drilling or welding could cause their weakening. High strength 1-K polyurethane and 1-K modified polymer adhesive systems were tested, four in total, in combination with different representatives of the substrate. Joint design similar to butt joint was used, therefore each test sample was composed of two components: (1) EN AW-6060 (AlMgSi) T66 aluminium alloy disc; and (2) the tested substrate: (a) COR-TEN® steel, (b) aluminium composite panel (Alu-Bond), and (c) high-pressure laminate (HPL). Moreover, all test samples were exposed to artificial weathering, as a result the effect of different temperatures (ranging from −20 up to +70 °C) and environments (from 0 up to 100% humidity) on bond strength and failure mode was monitored. The relation between bond failure energy density and bond tensile strength was investigated. For all the test samples a consistent relation was observed, however, the strain energy density that corresponds to the bond failure energy was entirely different for each of the adhesives tested. The dependency of adhesive stiffness on strain energy density was manifested. On the contrary, the effect of different weathering methods was not dominant, only slight deviation were observed in the combination with HPL substrate.

Electronic supplementary material The online version of this chapter (https://doi.org/10.1007/978-981-15-6767-4_6) contains supplementary material, which is available to authorized users.

B. Nečasová (✉) · P. Liška · J. Šlanhof
Brno University of Technology, Veveří 331/95, 602 00 Brno, Czech Republic
e-mail: necasova.b@fce.vutbr.cz

Lucas F. M. da Silva et al. (eds.), *Industrial Applications of Adhesives*, Lecture Notes in Mechanical Engineering,
https://doi.org/10.1007/978-981-15-6767-4_6

Keywords Adhesive · Adhesion · Bond · Cladding · Composite · Deformation · Energy · Façade · Joint

1 Introduction

The use of adhesives in various fields of the industry represents a considerable challenge as the ongoing supply of materials on the market is very broad, almost inexhaustible, in an effort to satisfy requirements of all investors. Innovative lightweight and composite materials often with mischievous surface properties are currently the most demanded alternatives not only in the industrial sector of the engineering industry. The use of e.g. high strength steel, aluminium alloys or recyclable composites is no more a novelty, and without any doubts, it can be stated that these materials are nowadays the most commonly used variants across all industries including the civil, automotive, aeronautic, railway, marine or energy sectors [1–5].

In construction applications, mainly in façade applications that are addressed in the presented paper, the joining of various parts and components is usually ensured by traditional joining methods [6, 7]. These are mechanical joints; made by nailing, riveting or screwing; and welded joints. Efficiency and durability of these fastening methods was already proven many times, moreover, they offer reasonable price and simpler implementation requirements [4]. On the other hand, both methods still suffer from certain limitations and some recent studies indicate that they cause a higher failure rates compared to adhesive joints [1, 3, 4, 8–11]. As stated by de Freitas et al. [1] or Ozel et al. [4], to overcome these limitations adhesive bonding ought to be considered. Besides, adhesive joints offer benefits as is the implementation of permanently flexible joints with uniform stress distribution, avoidance of pre-drilling or in-situ drilling necessity with subsequent appearance of local stress concentration [12]. Many authors also agree than adhesively bonded joints are the most suitable technology for joining of composite materials since many adhesives were developed in parallel with composites [1, 3, 13, 14].

The presented case study was initiated by a market representative. The main objective is to analyse the deformation behaviour and tensile strength of adhesively bonded joints made of aluminum alloy in combination with different adherends, namely: aluminum composite panel (ACP: ALU-bond), weathering steel (CORTEN) and high-pressure laminate (HPL), (author's note: the selection of the aforementioned materials was carried out by the market representative). To specify the deformation behaviour and bond strength the effect of variable temperature changes was considered, as the tested assembly is installed mostly in the outdoor environment. Moreover, the effect of temperature was verified since it has been already acknowledged by many authors that temperature changes are one of the most problematic factors when it comes to the durability of adhesive joints [2, 5, 15–25]. Furthermore, one of the client's main inquiries was: '*How much the changes in outside temperature will affect the strength and durability of the façade assembly, and at the same time, whether any of the selected adhesive systems will exhibit different deformation and failure behaviour if it is regularly exposed to temperatures higher than are the current norms?*', since structures are commonly constructed for periods longer than 50 or even 80 years. The standard temperature range in the Czech Republic used to be −5 up to 30 °C, however, during the last 5 years the temperature usually ranges in −20 to +40 °C. It is known that ideal results of adhesive joints can be obtained in the temperature range for which the adhesive has been designed [25], however, the climate changes, that we are recently experiencing, results in a change of habitat conditions (e.g. temperature increase and/or decrease, change in precipitation distribution etc.) that can directly affect the durability of currently constructed structures.

The effect of temperature on the durability and deformation behaviour of the adhesive joints with aluminium, steel and composite materials, mainly with carbon based composites, has lately earned the attention of many authors. Ozel et al. [4] studied the strength of adhesively bonded joints with aluminum and carbon/epoxy composite adherends; and Rahmani et al. [5] analysed the thermo-fracture behaviour of composite-aluminum joints at low temperatures; both studies centred on the aeronautic sector; Avendaño et al. [2] verified the effect of temperature and strain rate on single lap joints with unidirectional carbon fibre reinforced polymer; and Na et al. [21] verified the effect of temperature on the mechanical properties of adhesively bonded basalt FRP-aluminum alloy joints; both works were focused on applications in automotive industry. The influence of elevated temperatures on epoxy adhesive used in CFRP strengthening systems for civil engineering applications was recently described by Firmo et al. [22]. Unfortunately, only a small number of the published works focus on thick layer adhesive joints that are used in façade applications. In 2010, Banea et al. [15] confirmed that increasing temperature has a negative effect on strength and stiffness of 2 mm thick bulk specimens made of polyurethane adhesive, but does not affect the strain properties. Moreover, they confirmed that ambient temperature ranging in −5 up to 30 °C has generally very small influence on bond strength; on the other hand, it was proved that temperature lower or higher than the specified range has a prominent effect. Na et al. [20] obtained similar results in combination with silyl modified polymer based adhesive in 2017, and confirmed the temperature dependency of failure behaviour, however, unlike the polyurethane adhesive tested by Banea et al. [15], an improvement of the stress/strain behaviour was monitored if the temperature decreased. In both case studies, the presented results confirmed that temperatures above Tg are critical for bond stiffness and durability. In 2019 Machalická et al. [6] verified that mechanical properties of 5 mm thick silane terminated polymer (STP) adhesive, that can be used also in façade applications, are highly influenced by environmental conditions; 12% strength reduction was monitored. The use of long thick-layer bonds in façade applications was also verified by Bues et al. [7] in 2019. The strength and failure behaviour of two thicknesses of 2 K silicone materials (i.e. 6 and 12 mm) was described. They observed a behaviour identical to the one presented by Banea et al. [15] and Machalická et al. [6], the temperature reduction led to an increase of the maximum load and elongation of the joint at break, whereas an increase of the temperature led to a strength reduction [7]. In all presented case studies aluminium, steel or composite substrates were tested. Even though some minor issues with adhesion were monitored, mainly by Machalická et al., the biggest problem was the exposure of the joint to a different environmental condition.

Hence, thermal cycling to verify the effect of variable temperatures on bond strength and failure behaviour is studied in the presented paper. The adhesives investigated in the presented paper find their main application in the construction industry, mainly in the area of façade applications. The adhesives described as Type I, Type II and Type III are part of mounting systems intended for façade applications. On the contrary, the adhesive referred to as a Type IV is a commonly available structural adhesive with general purpose of use.

2 Experimental Details

2.1 Materials

When choosing an adhesive for façade applications, it is important to consider not only its maximum strength but also its ductility and toughness since the materials to be joined usually have different deformation properties which can often cause premature failure of the joint.

2.1.1 Adhesives

The adhesives investigated in the presented paper find their main application in the area of façades, namely Type I, Type II and Type III. Moreover, the Type I and Type III are the only representatives recommended by the manufacturers for the bonding of ALU-bond and HPL substrates. The Type I and Type II are 1 K polyurethane moisture-curing permanently elastic adhesives. Type III and Type IV are representatives of 1 K MS polymer moisture-curing permanently elastic adhesives. Both groups of polymers have nowadays already a solid position on the market. Additionally, Type IV is commercially available structural adhesive, and also the only representative that does not require the use of surface activators-primers or surface roughening before the adhesive application. The relevant material properties of selected adhesives are presented in Table 1.

2.1.2 Adherends

The test specimen was composed of two parts. The first part was always made of the aluminium alloy EN AW-6060 (AlMgSi) T66 with a tensile strength of 195.0 MPa,

Table 1 Material properties of adhesive systems

Polymer-base	Polyurethane		MS Polymer	
Adhesive system	Type I	Type II	Type III	Type IV
Tensile strength (MPa)	2.50	9.00	2.50	3.50
Elongation at break (%)	480	approx. 700	approx. 400	approx. 200
Service temperature (°C)	−40 up to +150	−40 up to +100	−40 up to +100	−40 up to +120
Glass transition temperature (°C)	−40			

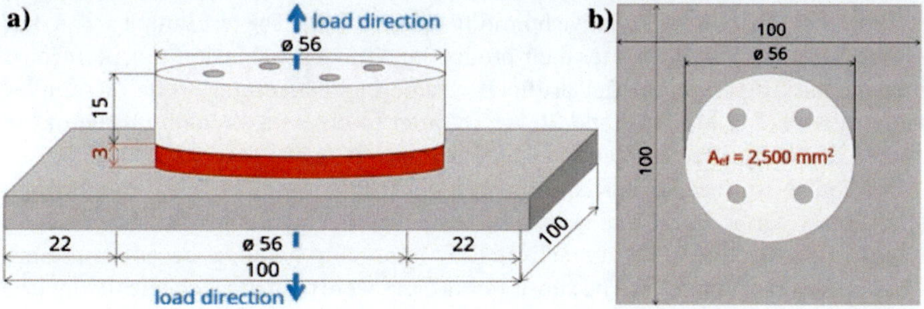

Fig. 1 Joint design: **a** axonometric and **b** top view of tensile test specimen in mm

a yield strength of 150.0 MPa and thickness 15 mm. This part of the test sample as well as the selected material substitute the load-bearing substructure of the facade system. The specimens were cut from the alloy plates in the form of circular discs of c. 56 mm in diameter, see Fig. 1.

The second part of the test sample was made of three different substrates that were selected by the market representative. All tested substrates were provided in a form of large format panels that were cut into samples with cross section dimensions l = b = 100 mm. The first tested substrate was aluminum composite panel (ALU-bond). The panel is a composite consisting of two thin layers of aluminum sheet made of EN AW-5005a (AlMg), with maximum 0.5 mm thickness, sandwiching a fire rated core. The core is made of LDPE. The outer aluminum skin was coated with fluorocarbon coating. The lower aluminum skin was chromate treated and polyester coated. The tensile strength of the material in board plane is 130.0 MPa. Panels 6 mm thick were used. The second representative of façade cladding was weathering corrosion resistant steel sheet (COR-TEN). The hot rolled 2 mm thick COR-TEN type A with tensile strength in board plane 485.0 MPa and yield strength 345.0 MPa was chosen. The third material tested was Trespa Meteon board that belong to the category of high pressure compact laminates (HPL). The tested product is made of 70 wt% wood fibres which are at high temperature and pressure saturated with 30 wt% epoxy resin. Boards 6 mm thick with tensile strength in board plane around 120.0 MPa were used.

2.2 *Joint Design*

The joint design was based on a recommendation of ČSN 73 2577 standard to obtain the tensile properties of the bonded assemblies. It is a design similar to the butt joint assembly. The technical standard is a national alternative to pull-off adhesion testing. The aluminium discs were bonded to selected substrates via application of 3 mm thick adhesive layer as shown in Fig. 1. Thickness of 3 mm is recommended for the tested facade applications [26, 27]. All test samples were fabricated in clean environment [(20 ± 3) °C, relative humidity (55 ± 10) %]. The test samples were left to static curing for 14 days after their production. The test in tension was performed after the curing period or after artificial weathering. Therefore, 3 sets of samples were tested, i.e. at RT, −20 and 70 °C, in order to obtain a strength profile of the bonded assembly affected by variable temperature changes.

All contact surfaces of test specimens were treated by mechanical roughening. The contact surface of aluminium discs was abraded by Scotch Brite arch 7446 S/MED, degreased/cleaned and activated by improving liquid, if necessary, before the application of adhesive. The substrate surfaces were treated mechanically by P80 grid sandpaper. Microscopic examination of surface topography was not performed; however, all test samples were roughened by the same person to avoid any result distortion. In combination with adhesives intended for facade bonding (i.e. Type I, Type II and Type III) improving liquid must be applied on a bonded area at least 30 min before the adhesive is used. All cleaning and activation liquids were used in accordance with manufacturer's requirements.

2.3 *Artificial Weathering*

Two test methods were chosen to simulate the natural environmental conditions that are greatly affecting the physical and mechanical properties of bonded joints. As stated by Machalická [6], primarily the effect of the combination of temperature and moisture can cause a significant degradation of an adhesive joint, therefore, both presented methods combine the impact of these actions.

Firstly, the freeze-thaw cycles were considered since these are a common operating condition for outdoor structures that requires the attention [28]. Periodic freeze-thaw cycles were imposed on the test samples, each cycle last 24 h as shown in Fig. 2a. The samples were placed into a climatic chamber and left there for 18 h. The change from +20 °C to −20 °C took around 14 h, the speed of temperature decrease was set to a rate of 2.0 °C/h, avoiding a possible thermal shock. After 18 h the samples were immersed in a water bath with initial water temperature (20 ± 3) °C and same

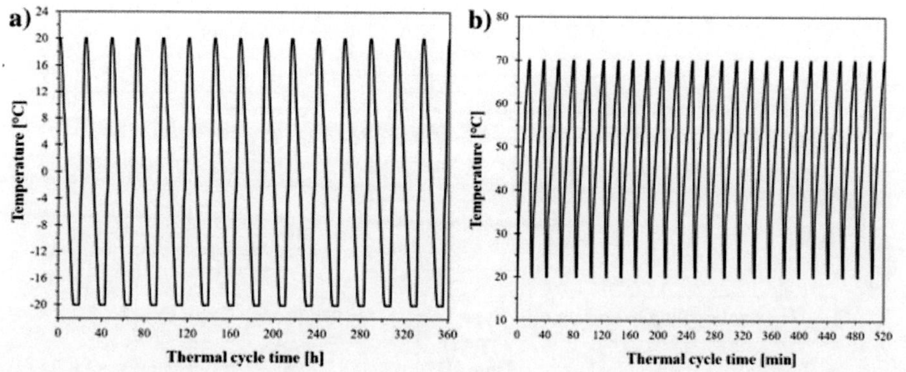

Fig. 2 Temperature profile of thermal cycling: **a** freeze-thaw cycles; **b** cycles of sudden temperature changes

environment temperature. The samples were left in the water bath for next 6 h. The change from −20 °C to +20 °C took around 4 h. The samples were imposed to 15 freeze-thaw cycles in total, which represents 360 h of conditioning. After the thermal cycling was ended the test samples were left to cure for 7 days.

The second weathering method is simulating the impact of sudden temperature changes. The test set-up composed of infra lamps and water bath was used, to at first heat the adhesive joint to a temperature of +70 °C. The temperature had to be reached in the bonded joint, not on the external surface of the substrate. Thermo imager was used to control the temperature. The infra lamps were placed 40 cm far from the external surface of the substrate. The temperature +70 °C was reached approximately after 15 min of exposure to the infra light. The test samples were subsequently cooled to a temperature of 20 °C by immersion to water bath. The change from +70 to +20 °C took around 5 min. The samples were imposed to 25 cycles. It was 524 min (i.e. c. 9 h) in total. The cooling process represented a significant thermal shock. After the cycling was ended the test samples were cured for 14 days.

2.4 *Tensile Test*

The samples were subjected to the displacement-controlled loading. The specimens were placed into a test apparatus that allowed application of controlled axial load. The displacement (i.e. the joint elongation) was measured by one inductive displacement transducers with a maximum deviation of 0.15% that was placed on the mobile (i.e. bottom) jaw of the testing machine. A Heckert FP 10/1 tearing machine with 10 kN capacity was used. The test samples were loaded continuously at speed of 5 mm/min until the break. The average value of tensile strength, stress and strain was calculated for each test set. The joint strength and temperature dependence were determined.

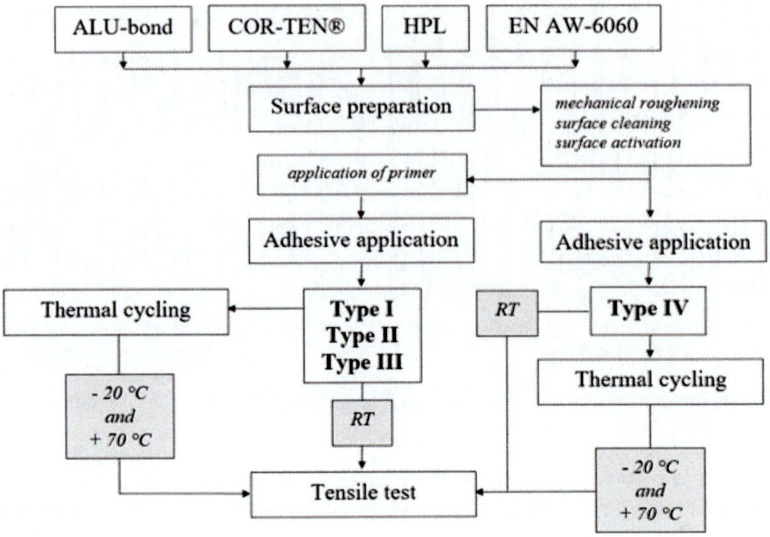

Fig. 3 Simplified flow diagram of the sample preparation for the tensile test

The average strain energy density was calculated for each adhesive. The steps of sample preparation from surface treatment to thermal cycling and tensile test are schematically summarized in Fig. 3.

3 Results and Discussion

In order to understand the influence of artificial thermal aging on deformation behaviour and tensile strength of bonded joints in combination with selected substrates a total of 216 samples were tested. The average values of strength and deformation properties are presented and discussed in this chapter. The maximum load was usually measured at the point of joint break. However, in combination with adhesive Type II the limit of the testing machine (i.e. 10 kN) was reached in some tested combinations of substrates, therefore, the presented values have not been measured at joint break as can be seen in Fig. 7b. Moreover, the maximum tensile strength of all joints was calculated and compared with the strength limits stipulated by the manufacturers. The adhesive Type I, irrespective of the bonded material, reached less than 70% of the strength, the Type II achieved 46% of the strength, for Type III 86% compliance with the maximum strength was achieved and for the Type IV, 79% of the strength stated by the adhesive manufacturer was reached. Furthermore, the tensile strength of the adhesive Type II should be around 9.0 MPa but in the presented paper, due to the limit of the test machine, only 50% of the maximum value stated by the manufacturer was reached. A minor defect of the joint was monitored only in 6 samples with adhesive Type II, while only slight dependence on the thermal cycling was observed.

The stress/strain relation of bonded assemblies was the main evaluation criterion. In 18 cases, the samples had to be excluded from the evaluation group and replaced by a new one. In these 18 cases, an extreme deviation from average values, that are presented in tables below, was observed. The maximum load measured at samples break was extremely low, usually around 1.5 kN ($\approx$0.6 N/mm^2). These extremes were observed only in combination with HPL substrate and adhesives Type I and Type II which are both 1 K polyurethanes. Even though the defect was observed predominantly after artificial weathering, a detailed inspection of the samples showed that the premature joint failure was not affected by thermal cycling. A colour indifference on the substrate surface was observed in all cases of the prematurely broken samples. The authors are of the opinion that this defect caused the premature failure of the joint since other divergences suggesting why the adhesive joint did not achieve the required strengths have not been monitored. The measured loads and elongations of these samples were only up to about 25% of the adhesive average. An adhesive failure in the interlayer between the substrate and the improving liquid was observed in all deteriorated samples. The thin layer of the primer always remained on the surface of the adhesive.

3.1 Effect of Thermal Cycling on Stress-Strain Behaviour

It was observed in the analysis that the mechanical properties of the selected adhesive systems were almost uninfluenced by tested environmental conditions. The stress-strain curves of all adhesive joints were recorded, as shown in Fig. 7. The values of tensile strength and failure strain presented in Tables in the supplement materials and in the graphical comparison in Figs. 4, 5 and 6 were calculated according to the recorded stress-strain curves for all tested material combinations. The obtained data were also used to calculate the failure energy (i.e. strain energy density).

3.1.1 Comparison of Stress-Strain Behaviour with ALU-Bond Substrate

As stated above, only minor effect of the different environmental conditions was observed in combination with ALU-bond substrate. It was expected that the tensile strength of all specimens will reduce after the artificial aging process, however, as can be seen in Fig. 4 only adhesives Type I and Type III were slightly affected by both conditioning methods.

The average tensile strength of samples with adhesive Type I after freeze-thaw cycles decreased by 2% and after the cycles of variable temperature changes by almost 6%. Moreover, the strain decreased as well and the effect of the high temperature on the reduction of the average strain was around 8%. This behaviour is quite typical

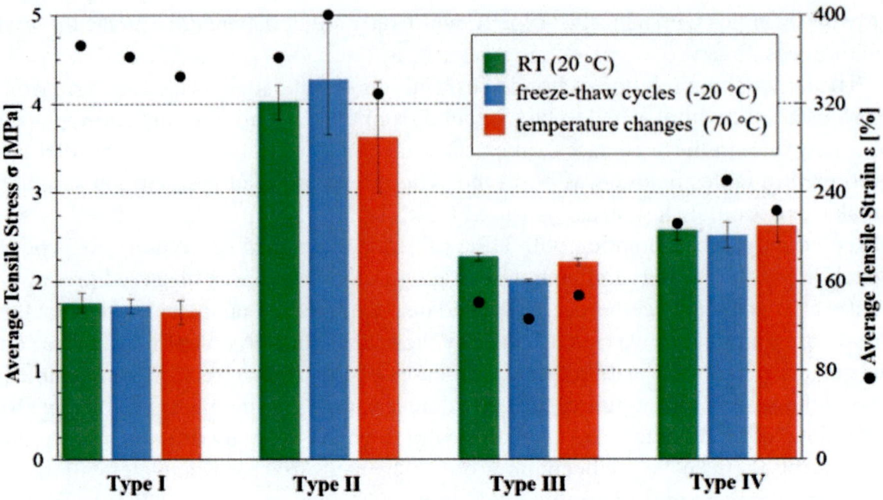

Fig. 4 The comparison of stress-strain behaviour with ALU-bond

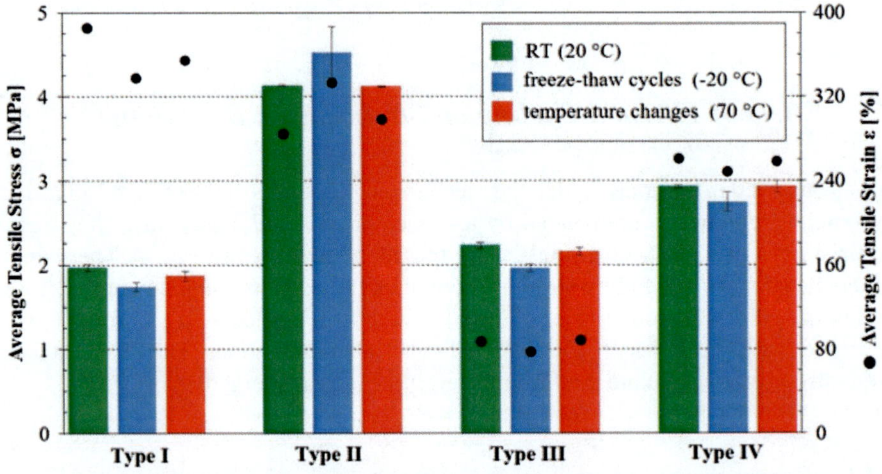

Fig. 5 The comparison of stress-strain behaviour with COR-TEN

for polyurethane based adhesives; however, the same pattern was not observed in combination with Type II. In this combination almost 10% decrease of tensile stress and strain was monitored only after the conditioning to sudden temperature changes. The results support one of the conclusion of Banea et al. [15], that mechanical properties of polyurethane adhesives tend to be negatively influenced by elevated temperatures more than low temperatures. The influence of elevated temperature was monitored also in combination with other tested substrates, see Figs. 5 and 6. On the contrary, the extent of variability of the measured data is too significant

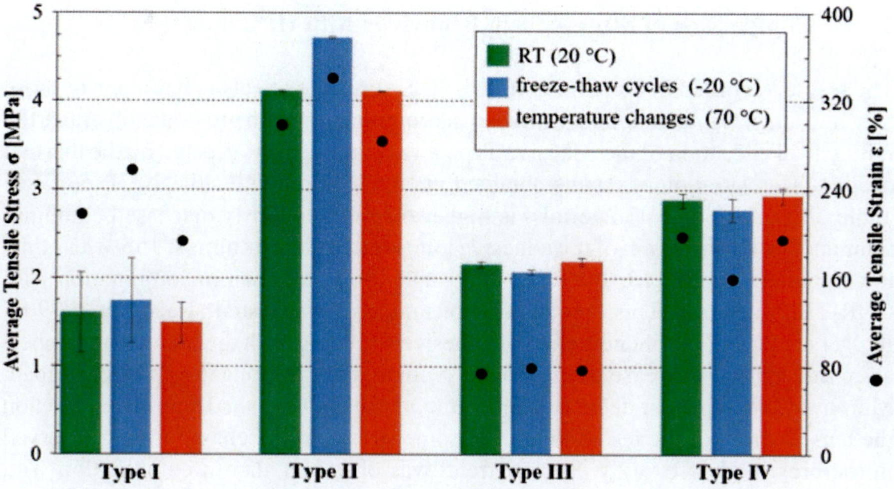

Fig. 6 The comparison of stress-strain behaviour with HPL

(see the highlighted values of CV in Table 1A in supplement materials) and further analysis of this effect is needed.

In combination with adhesive Type III the negative effect of temperature on stress-strain behaviour was monitored as well, however, unlike with Type IV, the conclusions presented by Na et al. [20] have not been confirmed, i.e. that the strain of MS polymer based adhesives increase with decreasing temperature.

3.1.2 Comparison of Stress-Strain Behaviour with COR-TEN Substrate

Similar stress-strain behaviour as in combination with ALU-bond substrate was observed in combination with COR-TEN steel. However, as can be seen in Fig. 5, the effect of temperature on mechanical properties of adhesive Type I was more severe after freeze-thaw cycles; with decreasing temperature a moderate decrease in the tensile strength occurred. The tensile strength of the adhesive joints tested at −20 °C is almost 13% less than in combination with ALU-bond substrate.

In combination with adhesive Type II the limit of testing machine was reached with almost all samples, therefore the 66% decrease of joint tensibility is unexpected. Similar results were determined for adhesive Type III. Even though the effect of artificial weathering on tensile stress is analogous with the results calculated for the combination with ALU-bond substrate, significant reduction of strain properties can be seen. The best results were observed in combination with adhesive Type IV. Compared to the results obtained in combination with ALU-bond substrate, an improvement of tensile strength as well as of the joint tensibility was verified. On the contrary, minor effect of temperature was confirmed.

3.1.3 Comparison of Stress-Strain Behaviour with HPL Substrate

The results recorded in combination with HPL substrate are also analogous to those presented in previous chapters. Unlike in the combination with previous substrates the failure load and strain of the adhesive Type I was significantly affected by the thermal cycling. The variation of results obtained in combination with adhesive Type I (see Table 3A in supplement materials) is higher than in previously discussed combinations and the tensile strain of the adhesive joints tested at RT is almost 155% less than in combination with ALU-bond substrate, and 168% less than in combination with COR-TEN substrate. This combination of materials was tested already in 1999 by Kruger et al. [29], the phenomenon of adhesive softening with the increasing temperature that led to a decrease of the tensile strength and the elongation was described. Moreover, the impact of decreased temperature (i.e. −20 °C) had a positive effect on the tensile strain of the tested joint. The same stress-strain behaviour was observed in the presented case study. Similar trend was observed also in combination with adhesives Type II and Type III. This effect was monitored also in combination with adhesive Type IV, however, the impact of temperature −20 °C was exactly the opposite as in previous cases and almost 20% strain decrease was recorded. Therefore, the optimistic assumption stated above that lower temperatures can positively affect the tensile strain of all MS polymer based adhesives was refuted.

3.2 Effect of Artificial Aging on Bond Toughness and Failure Mode

Even though only a minor effect of different environmental conditions on adhesive joint strength in combination with selected substrates was observed; various deformation behaviour was monitored in all tested samples as can be seen in Fig. 7. The verification of the effect of artificial weathering methods on bond toughness and failure performance was provided.

The comparison of strength and elongation relation presented in Fig. 7 shows that each of the tested adhesives performed very uniformly in combination with selected substrates. A smooth transition from the region of the elastic deformation to the plastic deformation region was noted in the case of both polyurethane adhesives. The elongation of the adhesive Type II in the region of elastic deformation was 2× larger than in combination with the adhesive Type I, however, also the average and maximum joint strength was the double of the values measured in combination with adhesive Type I.

Comparable progress in the region of elastic deformation was observed for all tested adhesives and material combinations, however, in combination with polyurethanes, the yield point occurred later after the elasticity limit was reached than in combination with MS polymer-based adhesives. Moreover, a constant strain growth was observed in the case of PU-adhesives after the yield point was exceeded

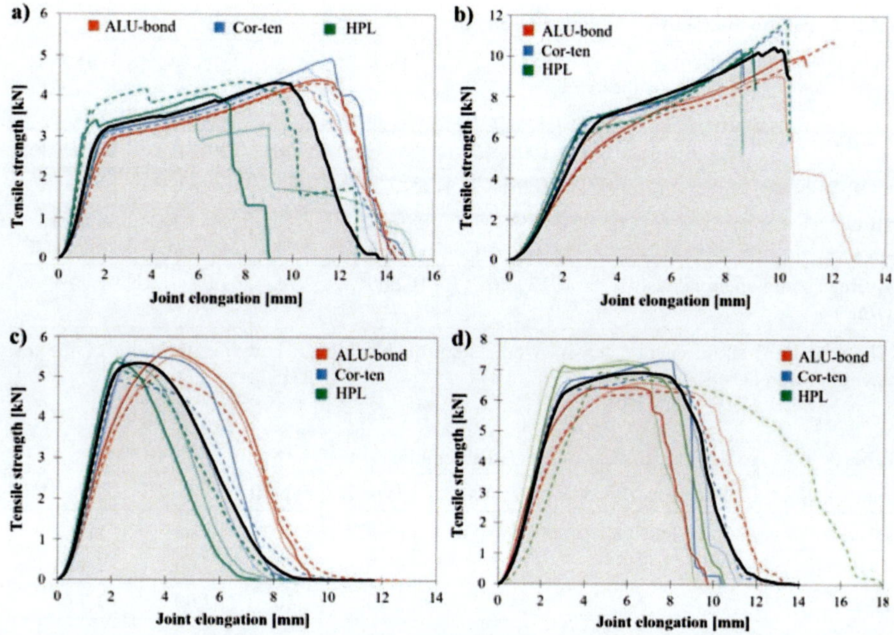

Fig. 7 Strength-elongation diagram: **a** Type I, **b** Type II, **c** Type III and **d** Type IV—Comparison of deformation behaviour based on average values of each tested set of specimens exposed to weathering conditions and of the reference deformation behaviour of the adhesive (solid black line); Solid lines are RT samples, dashed lines are −20 °C samples and dotted lines are +70 °C samples

as can be seen in Fig. 10 where the average failure mechanism of adhesive Type I is illustrated. This behaviour was not monitored in combination with MS polymers. The failure of the adhesive joint appeared soon after the yield point was reached. Due to the uniform deformation behaviour of all tested combinations a reference stress-strain curve for each of the tested adhesives was constructed and the reference strain failure energy density was calculated. According to Staudt et al. [30], the strain energy density can be used for the fatigue life prediction since the rupture of the material occurs, when the binding energy between molecules is overcame. Therefore, this energy corresponds to the strain energy density which is equal to the area below the stress-strain curve. The values presented in Table 2 are the calculated energy density needed to break the adhesive joint. Even though it is not the energy of the interaction at an interface, the authors are of the opinion that even these values can demonstrate the strength of the bond.

According to [31], the average energy necessary to break a polymer-based adhesive joint made at room temperature is on the order of 100 J/m^2. As can be seen in Table 2 the calculated values of the average strain energy density are very close to this standard. Moreover, on the basis of the author's experience, it was assumed that adhesive Type I is the ideal type of adhesive for façade application. The energy calculated

Table 2 Failure mechanism of bonded assemblies

Substrate	Failure property/adhesive system	Type I	Type II	Type III	Type IV
ALU-bond	Failure mode (–)	100% CF	18% CF; 82% NF	100% CF	100% CF
COR-TEN		100% CF	6% CF; 94% NF	100% CF	100% CF
HPL		60% CF 40% A/CF	88% NF 12% A/CF	76% CF 24% A/CF	88% CF 12% A/CF
Average strain energy density (J/m^2)		129.56	162.69[a]	61.07	91.90

[a]The calculated strain energy density would be probably higher, however, the limit of the test machine was reached

Table 3 Average tension stiffness of bonded assemblies

Substrate	Failure property/adhesive system	Type I	Type II	Type III	Type IV
ALU-bond	Average tension stiffness k (MPa)[a]	0.48	1.12	1.65	1.28
COR-TEN		0.53	1.46	2.59	1.14
HPL		0.79	1.46	2.99	1.77

[a]The tension stiffness of elastic support was determined. The values of 95% quantile of a random quantity are presented

for the adhesive Type I is very close to the above stated ideal and also the average tension stiffness presented, in Table 3, is the lowest from the tested combinations. Considering that the higher the value of the tension stiffness, the worse the properties of the adhesive joint/assembly, the abovementioned statement was confirmed.

The energy calculated for adhesive Type II and for both MS polymer-based adhesives corresponds to the observed stiffness. These adhesives were less ductile and stiffer which. This was also the main cause of their failure and the most common reason why smaller elongation of the joint at break was recorded.

On the contrary in all tested combinations the cohesive failure mode was the most frequent. In some cases, the combination of slight adhesive failure with cohesive failure was monitored, see examples in Figs. 8, 9 and 10. As can be seen in Fig. 7a, c the adhesive Type I and Type III showed a higher failure rate in combination with HPL substrate. The adhesive failure in combination with Type I was also the cause of the very high value of the coefficient of variance (see Table 1C in supplement materials).

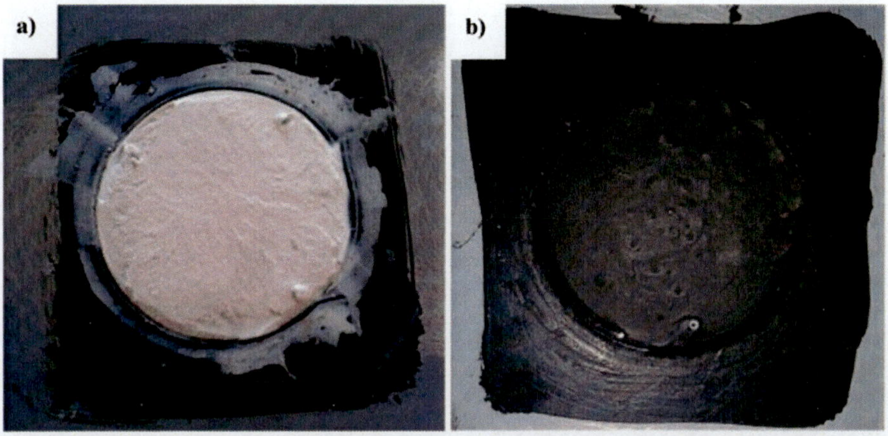

Fig. 8 Adhesive failure with ALU-bond substrate: **a** adhesive Type I; **b** adhesive Type II

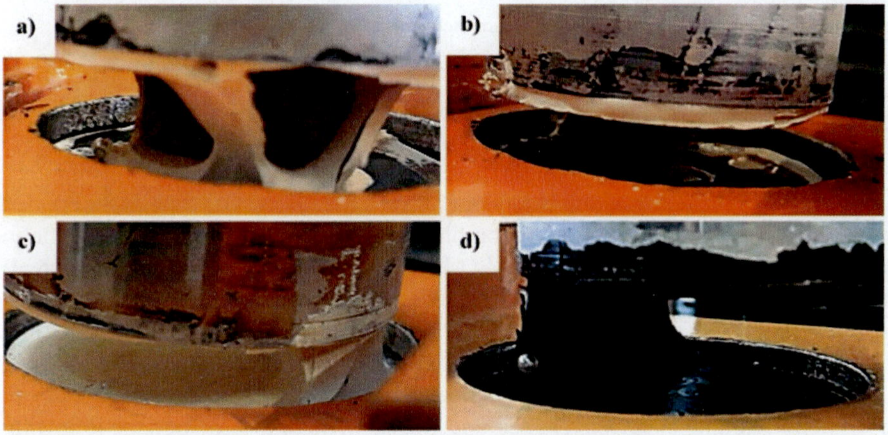

Fig. 9 Examples of different failure modes: **a** combination of adhesive and cohesive failure with adhesive Type I and HPL substrate; **b** combination of adhesive and cohesive failure with adhesive Type I and HPL substrate; **c** cohesive failure with adhesive Type I and ALU-bond substrate; **d** cohesive failure with adhesive Type II and ALU-bond substrate

4 Conclusion

The bonding ability of dissimilar adherends bonded with 1 K polyurethane or 1 K silyl modified polymer-based adhesives intended for façade applications was assessed via the tensile testing under axial loading. The quantitative description of stress-strain behaviour was recorded. Moreover, the effect of temperature on the deformation behaviour of the adhesive joint was evaluated. The impact of three different environment conditions was observed, i.e. of the ambient temperature (20 ± 3 °C); the

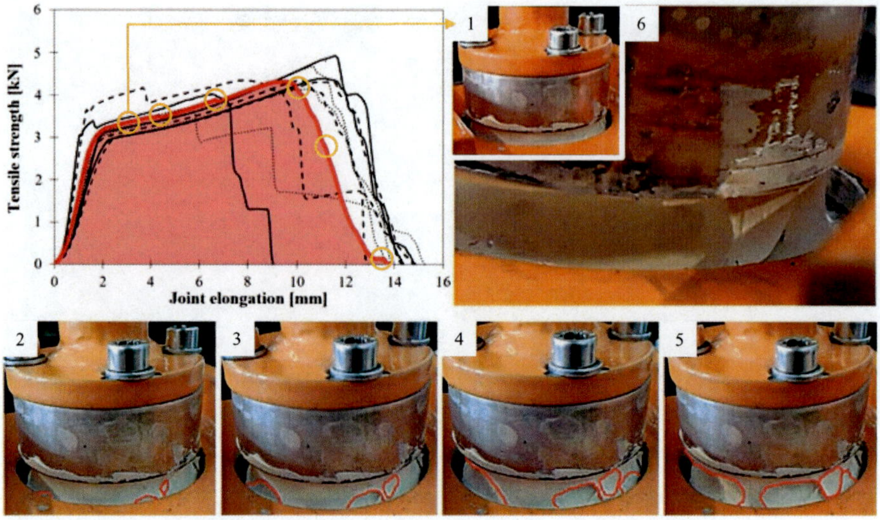

Fig. 10 The ilustrative example of the failure mechanism of adhesive Type I

freeze-thaw cycles (−20 ± 2 °C) and cycles of sudden temperature changes (70 ± 3 °C).

The presented case study was initiated by a market representative with an intention to analyse the deformation behaviour and tensile strength of adhesively bonded joints made of aluminium alloy in combination with different adherends, namely: aluminium composite panel (ACP: ALU-bond), weathering steel (COR-TEN steel) and high-pressure laminate (HPL). The secondary objective of this study was the price of tested adhesives. The average price of adhesive Type I and Type III is 2 times higher than the price of adhesives Type II and Type IV. Moreover, adhesive Type IV is not intended for façade applications, it is only a structural glue that can be used for bonding of selected adherends. Based on these information the following conclusions were summarized:

1. All tested substrates are well-suited for adhesive bonded. The analysis of results proved great compatibility with all selected adhesives. A reliable bond was achieved in all combinations.
2. It was observed that the adhesive Type I reached less than 70% of the tensile strength specified in the technical data sheet by its manufacturer. Moreover, the tensile strength and strain were slightly affected by both conditioning methods. With increasing temperature, the adhesive was more ductile and with a lower capacity to withstand higher loads. The adhesive Type II achieved only 46% of the stipulated maximum strength. However, this result was greatly influenced by reaching the limit load of the test machine. After reaching the maximum load that the test equipment could exert (10 kN), the test machine stopped. For the adhesive Type III almost 89% of the predefined tensile strength was achieved and thermal dependency was monitored. The adhesive affected by thermal cycles was less

ductile and also less durable. In combination with Type IV the average strength was around 79% and thermal sensitivity was not confirmed.

3. Only three different failure modes were observed. The cohesive failure of the joint was monitored in 88 and 6% of samples broke by the combination of adhesive and cohesive failure, where the cohesive failure was always preponderate. The adhesive failure was monitored as well but it was mainly in combination of PU-based adhesives and HPL substrate, nevertheless, this anomaly was caused by deteriorated surface properties of the tested substrate. The defective samples were excluded from the evaluation.
4. The calculated values of average strain energy density are closely related to the calculated values of joint stiffness. The stiffer the joint, the lower strain was recorded.
5. The greatest strength was achieved with PU-based adhesive Type II, it was on an average of 10.45 kN. The second highest strength was recorded with the MS polymer-based adhesive Type IV. It was on the average of 6.92 kN. It has to be emphasized that it is the universal structural adhesive which is not primarily intended for bonding of façade elements, unlike the other tested adhesives; yet its deformation behaviour was less sensitive to thermal cycles. Moreover, the bonded substrate was activated only by a general cleaner. Any primer was not used. The use of multi-purpose structural adhesives in façade applications is quite common, however, if a failure occurs, its impact is usually much worse since only certified systems are recommended for this application sphere.
6. In the future work, a more diverse range of temperatures should be studied. The impact of temperature below the glass transition temperature (−40 °C) on the tensile strength must be observed. Moreover, the effect of elevated temperature above the service temperature range might be studied (100–150 °C). The finite element model will be included to simulate the deformation behaviour.

Acknowledgements The research case presented in this paper was supported by Brno University of Technology Internal Grant Programme project No. FAST-S-19-5805.

References

1. de Freitas ST, Sinke J (2015) Failure analysis of adhesively-bonded skin-to-stiffener joints: Metal–metal vs. composite–metal. Eng Fail Anal 56:2–13. https://doi.org/10.1016/j.engfailanal.2015.05.023
2. Avendaño R, Carbas RJC, Marques EAS et al (2016) Effect of temperature and strain rate on single lap joints with dissimilar lightweight adherends bonded with an acrylic adhesive. Compos Struct 152:34–44. https://doi.org/10.1016/j.compstruct.2016.05.034
3. de Freitas ST, Banea MD, Budhe S et al (2017) Interface adhesion assessment of composite-to-metal bonded joints under salt spray conditions using peel tests. Compos Struct 164:68–75. https://doi.org/10.1016/j.compstruct.2016.12.058

4. Ozel A, Yazici B, Akpinar S (2014) A study on the strength of adhesively bonded joints with different adherends. Compos B Eng 62:167–174. https://doi.org/10.1016/j.compositesb.2014.03.001
5. Rahmani A, Choupani N, Kurtaran H (2019) Thermo-fracture analysis of composite-aluminum bonded joints at low temperatures: experimental and numerical analyses. Int J Adhes Adhes 95:102422. https://doi.org/10.1016/j.ijadhadh.2019.102422
6. Machalická K, Vokáč M, Kostelecká M et al (2019) Structural behavior of double-lap shear joints with metal substrates under humid conditions. Int J Mech Mater Des 15:61–76. https://doi.org/10.1007/s10999-018-9404-y
7. Bues M, Schuler C, Albiez M et al (2019) Load bearing and failure behaviour of adhesively bonded glass-metal joints in façade structures. J Adhes 95:653–674. https://doi.org/10.1080/00218464.2019.1570158
8. Friedrich D, Luible A (2015) Measuring the wind suction capacity of plastics-based cladding using foil bag tests: a comparative study. J Bulid Eng 8:152–161. https://doi.org/10.1016/j.jobe.2016.10.009
9. Cheng R, Zhang L, Li Y (2012) Study on bonding properties of PVC-based WPC bonded with acrylic adhesive. J Adhes Sci Technol 26:2729–2735. https://doi.org/10.1080/01694243.2012.701477
10. Soury E, Behravesh AH, Rouhani Esfahani E et al (2009) Design, optimization and manufacturing of wood-plastic composite pallet. Mater Des 30:4183–4191. https://doi.org/10.1016/j.matdes.2009.04.035
11. Mohamadzadeh A, Rostampour Haftkhani A, Ebrahimi G et al (2012) Numerical and experimental failure analysis of screwed single shear joints in wood plastic composite. Mater Des 35:404–413. https://doi.org/10.1016/j.matdes.2011.09.061
12. Rodríguez RQ, de Paiva WP, Sollero P et al (2012) Failure criteria for adhesively bonded joints. Int J Adhes Adhes 37:26–36. https://doi.org/10.1016/j.ijadhadh.2012.01.009
13. da Costa Mattos HS, Monteiro AH, Palazzetti R (2012) Failure analysis of adhesively bonded joints in composite materials. Mater Des 33:242–247. https://doi.org/10.1016/j.matdes.2011.07.031
14. Budhe S, Banea MD, de Barros S et al (2017) An updated review of adhesively bonded joints in composite materials. Int J Adhes Adhes 72:30–42. https://doi.org/10.1016/j.ijadhadh.2016.10.010
15. Banea MD, da Silva LFM (2010) The effect of temperature on the mechanical properties of adhesives for the automotive industry. Proc Inst Mech Eng Part L: J Mater: Des Appl 224(2):51–62. https://doi.org/10.1243/14644207JMDA283
16. Banea MD, de Sousa FSM, da Silva LFM et al (2012) Effects of temperature and loading rate on the mechanical properties of a high temperature epoxy adhesive. J Adhes Sci Technol 25(18):2461–2474. https://doi.org/10.1163/016942411X580144
17. Banea MD, da Silva LFM, Campilho RDSG (2012) Effect of temperature on tensile strength and mode I fracture toughness of a high temperature epoxy adhesive. J Adhes Sci Technol 26:939–953. https://doi.org/10.1163/156856111X593649
18. da Silva LFM, Adams RD (2007) Adhesive joints at high and low temperatures using similar and dissimilar adherends and dual adhesives. Int J Adhes Adhes 27(3):216–226. https://doi.org/10.1016/j.ijadhadh.2006.04.002
19. Grant LDR, Adams RD, da Silva LFM (2009) Effect of the temperature on the strength of adhesively bonded single lap and T joints for the automotive industry. Int J Adhes Adhes 29(5):535–542. https://doi.org/10.1016/j.ijadhadh.2009.01.002
20. Na J, Liu Y, Fan Y (2016) Effect of temperature on the joint strength of a silyl modified polymer based adhesive. J Adhes 93(8):626–639. https://doi.org/10.1080/00218464.2015.1128330
21. Na J, Mu W, Qin G et al (2018) Effect of temperature on the mechanical properties of adhesively bonded basalt FRP-aluminum alloy joints in the automotive industry. Int J Adhes Adhes 85:138–148. https://doi.org/10.1016/j.ijadhadh.2018.05.027
22. Firmo JP, Roquette JR, Correia JR et al (2019) Influence of elevated temperatures on epoxy adhesive used in CFRP strengthening systems for civil engineering applications. Int J Adhes Adhes 93:8–18. https://doi.org/10.1016/j.ijadhadh.2019.01.027

23. Zhang J, Cheng X, Guo X et al (2020) Effect of environment conditions on adhesive properties and material selection in composite bonded joints. Int J Adhes Adhes 96:102302. https://doi.org/10.1016/j.ijadhadh.2018.12.001
24. Goda Y, Sawa T (2011) Study on the effect of strain rate of adhesive material on the stress state in adhesive joints. J Adhes 87(7–8):766–779. https://doi.org/10.1080/00218464.2011.597308
25. Adams RD, Coppendale J, Mallick V et al (1992) The effect of temperature on the strength of adhesive joints. Int J Adhes Adhes 12(3):185–190. https://doi.org/10.1016/0143-7496(92)90052-W
26. Dukes WA, Bryant RW (1969) The effect of adhesive thickness on joint strength. J Adhes 1:48–53. https://doi.org/10.1080/00218466908077374
27. Davies P, Sohier L, Cognard JY (2009) Influence of adhesive bond line thickness on joint strength. Int J Adhes Adhes 29:724–736. https://doi.org/10.1016/j.ijadhadh.2009.03.002
28. Yang Y, Silva MAG, Biscaia H et al (2019) Bond durability of CFRP laminates-to-steel joints subjected to freeze-thaw. Compos Struct 212:243–258. https://doi.org/10.1016/j.compstruct.2019.01.016
29. Krüger G, Schneider R (1999) An elastic adhesion system for structural bonding of facade panels. Otto-Graf-J 10:87–98. https://www.mpa.uni-stuttgart.de/en/institute/publications/otto-graf-journal/
30. Staudt Y, Odenbreit C, Schneider J (2018) Failure behaviour of silicone adhesive in bonded connections with simple geometry. Int J Adhes Adhes 82:126–138. https://doi.org/10.1016/j.ijadhadh.2017.12.015
31. Mark JE (2007) Physical properties of polymers handbook, 2nd edn. Springer Science & Business Media, Berlin, p 1076. ISBN 9780387690025

Ageing Condition Determination of Bonded Joints by Terahertz Spectroscopy

Jochen Taiber, Martin Kahlmeyer, Andreas Winkel, Eva-Maria Stübling, Fatima Taleb, Mikhail Mikerov, Stefan Sommer, Stefan Böhm, and Martin Koch

Abstract The understanding of ageing processes in the field of adhesion technology is of utmost importance for long-term durable bonded joints. Environmental factors such as moisture, heat and UV radiation negatively affect the characteristics of adherends, adhesives and their interfaces and thus the overall bond performance over time. However, the evaluation of the precise ageing condition of adhesive joints remains challenging by means of well-established non-destructive testing (NDT) methods like shearography or thermography. Recently, it has been shown that terahertz (THz) time-domain spectroscopy is well suited as an NDT approach to study adhesive bonds. Specialized data extraction schemes enable the contact-free determination of the refractive index and the absorption coefficient from such measurements. The presented work wants to give an overview on the developments in the application of THz spectroscopy for the ageing condition determination of bonded joints, including the correlation between THz refractive index/absorption and the physical, chemical and mechanical property changes of adherends and adhesives due to ageing, the determination of THz specific values of adhesive joints by means of a reflection setup, the suitability of peak-to-peak amplitude evaluation, and THz imaging.

Keywords Non-destructive testing · Terahertz spectroscopy · Ageing determination

1 Introduction

The importance of adhesive bonding is still ever-growing, favoured among other important joining technologies, due to the high demand of new innovative products and novel manufacturing processes [1].

J. Taiber (✉) · E.-M. Stübling · F. Taleb · M. Mikerov · S. Sommer · M. Koch
Department of Physics, Philipps University of Marburg, Marburg, Germany
e-mail: jochen.taiber@physik.uni-marburg.de

M. Kahlmeyer · A. Winkel · S. Böhm
Chair for Cutting and Joining Technologies, University of Kassel, Kassel, Germany

Lucas F. M. da Silva et al. (eds.), *Industrial Applications of Adhesives*, Lecture Notes in Mechanical Engineering,
https://doi.org/10.1007/978-981-15-6767-4_7

Especially for structural, and thus safety relevant, bonds, the reliability of the joints must be guaranteed for the intended lifetime. For this reason, a fundamental understanding of the ageing processes of adhesives and their bonds is also necessary, as well as a technology that can distinguish different ageing degrees of substrates, adhesives, and interfaces. First, promising approaches exist for validation of processes like the adhesive's application, in which bonds are examined for inhomogeneities like inclusions, pores etc. using imaging techniques [2]. Thermography [3] and shearography [4] should be mentioned here.

Currently, however, there is a high demand for technologies to reliably evaluate the quality or performance of an adhesive bond (the so-called "ageing condition" [5]) even after years of use in the field. Terahertz (THz) spectroscopy and imaging [6, 7] can provide an applicable approach to indicate non-destructively ageing processes within adhesive bonds [8, 9]. The authors would like to give an overview of different THz methods which enable the determination of ageing conditions of these joints.

2 Terahertz Time-Domain Spectroscopy and Imaging

Terahertz radiation is electromagnetic radiation in the wavelength range from 0.03 to 3.0 mm or a frequency of 100–10,000 GHz (0.1–10 THz). In a THz time-domain spectroscopy setup (THz TDS), this frequency range is often generated and detected by photoconductive antennas: A laser beam, which is divided into an emitter and a detector path, is initially focused on a semiconductor (LT-GaAs) and thus generates free charge carriers, which are accelerated by an applied voltage, resulting in emitted THz radiation (emitter path). After interaction with the sample, the THz radiation hits the detector antenna, which is enabled by the second part of the laser beam (detector path) to detect radiation in form of a photocurrent. The temporal sampling of the signal takes place via a delay unit typically integrated in the detector path, shortening and lengthening it (Fig. 1). For a more detailed description of the THz TDS technology see [6, 10].

Based on the time delay (shift in time) and the change in amplitude of the sample signal compared to a reference pulse, the determination of optical or dielectric properties such as the frequency-dependent refractive index n and the frequency-dependent

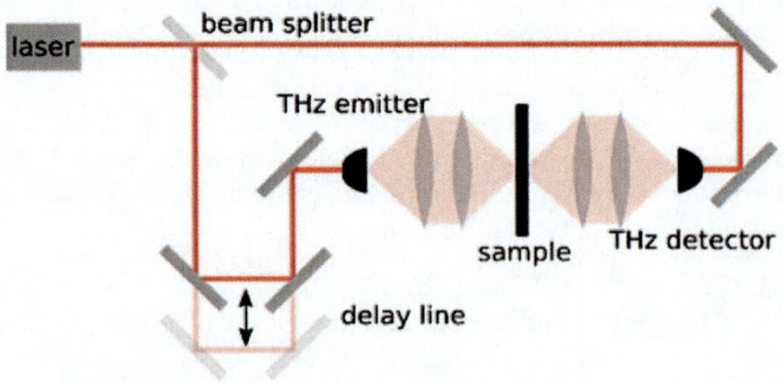

Fig. 1 Scheme of a THz spectrometer

absorption coefficient α becomes possible. In the present case, the analysis is done by an iterative minimization algorithm as described by Scheller et al. [11].

THz sensor technology now covers a wide range of application area, e.g. as an imaging method for security checks at airports or the screening of production goods for damage or foreign objects [12]. Furthermore, chemical substances can be identified by THz absorption spectra [13–16]. In addition, layer thicknesses [17] as well as the glass transition region of plastics [18] or the cross-linking of adhesives [19] can be monitored. The THz TDS technique can also provide information about the crystal structure of polymers [20, 21]. Even ageing effects of some polymers can be investigated [22, 23]. Besides polymers, most dielectrics can be studied using THz spectroscopy. This includes undoped semiconductors [24], glasses [25] and molecular crystals [26]. Materials containing water are transparent only to a small extend [27]. Metals reflect THz waves, only very thin layers can be penetrated.

In the following, various approaches for obtaining information on the ageing condition of adhesive bonds and materials in general will be presented.

3 Terahertz Spectroscopy for Determining Ageing Conditions

3.1 Ageing Determination of Polymers (adhesives and Substrates) in Transmission

In the present case, a commercially available time-domain spectrometer "T-Spectralyzer" (Hübner, Germany) was used. To allow for a vertical incidence of the THz radiation on the sample (not possible with the inline system), the 1550 nm laser beam was decoupled from the system by optical fibers and the THz path was set up freely on an optical table using HDPE lenses. A surrounding box filled with nitrogen excluded the disturbing influence of the air humidity. Different positions on the samples could be measured by using a focused spot and a mechanical linear unit.

For the transmission experiments, five measurements each were performed on two test specimens at three different positions each. In order to generate a sufficient signal-to-noise ratio, measurement data were recorded over a period of 30 s. The error margins and error bars, respectively, indicate the standard error determined from the individual measurements.

Figure 2 shows the frequency-resolved absorption coefficient and refractive index of the 2C epoxy adhesive 3 M Scotch-Weld DP100 from 0.3 THz to 2.0 THz. Further adhesives of different chemical bases were analyzed in [28]. To compare differently aged materials (Figs. 3 and 9), the refractive index or absorption coefficient at a frequency of 1 THz is used in the following, since none of the examined samples shows characteristic absorption lines in the frequency range considered here.

We investigated different ageing influences: One sample series was stored in water at 50 °C, another was heated up to 120 °C and a third was exposed to UV radiation.

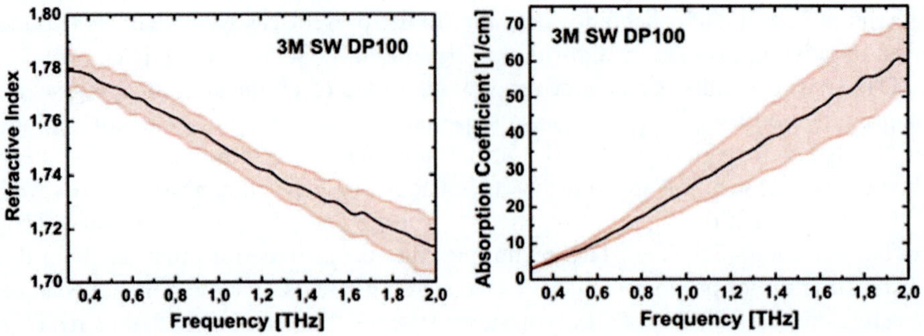

Fig. 2 Example of the frequency-dependent refractive index and absorption coefficient for the 2C-epoxy 3 M DP 100

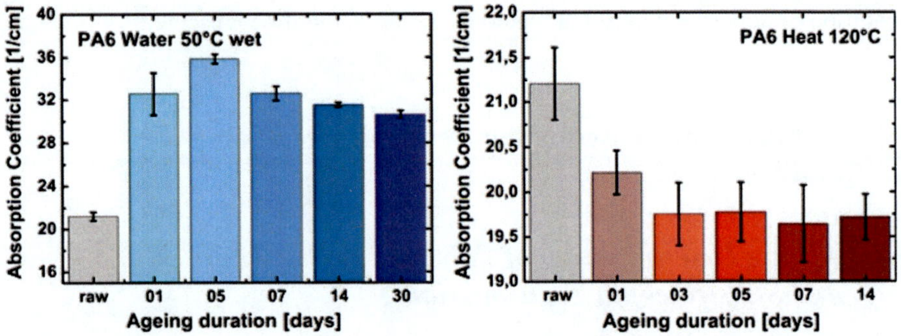

Fig. 3 Absorption coefficient of PA6 at 1 THz after water immersion (wet samples) and heat treatment at 120 °C

Figure 3 shows an example of the development of the absorption coefficient of polyamide 6 (PA6) as commonly used substrate after water immersion (wet sample measured) and after heat treatment for different time periods. It can be clearly seen that the refractive index rises sharply with water immersion, only to decrease slightly with increasing duration. Also, very clearly visible is the continuous decrease in the absorption coefficient of PA6 with increasing storage time of the samples at 120 °C. PA6, as a very polar polymer, is able to absorb about 8–9 mass percentage of water. Therefore, the observed dynamics can be directly associated with an increasing or decreasing water content, since water absorbs THz radiation very strongly.

This water absorption or release is also directly reflected in the mechanical and physical-chemical characteristics of the material. Examples are Young's modulus and glass transition (Fig. 4). The glass transition temperature was determined by torsional DMTA measurements (MCR 502, Anton Paar Germany) at a constant heating rate of 2 K/min and a linearly varying deflection angle (0.01–0.5%). The glass transition temperature was interpreted as the maximum of the loss factor (tan δ). The Young's modulus was characterized in accordance to DIN EN ISO 527 (type *1BA*) with a

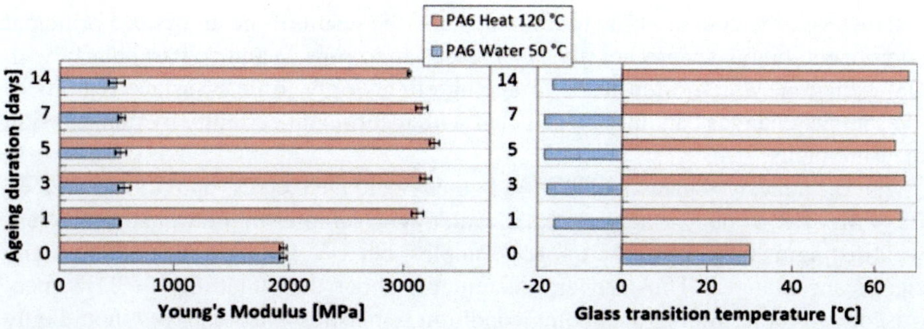

Fig. 4 Young's modulus and glass transition temperature for PA6 after water immersion and heat treatment

cross-head speed of 10 mm/min in the elongation range between 0.05 and 0.25% (Z100, Zwick-Roell Germany).

In order to be able to compare spectra as a whole more easily, they can be reduced to a few relevant components by means of principal component analysis [29]. This method is also used in FT-IR and Raman spectroscopy, for example, to facilitate direct comparison of a large number of samples. In the case of PA6, clearly defined clusters are formed (Fig. 5) which allows for a distinction of the various ageing influences.

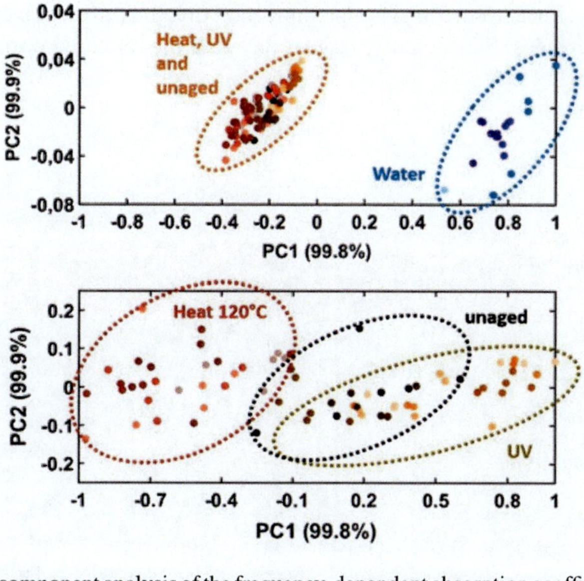

Fig. 5 Principal component analysis of the frequency-dependent absorption coefficients of different ageing methods and intensities of PA6. Each point represents a sample spectrum. In the upper picture all ageing methods are shown, in the lower the unaged samples and those after heat and UV influence in detail

However, the course of the measurands and the result of the associated principal component analysis were not always as clear as in Figs. 3 and 5. Particularly with UV radiation, which often leads to ageing effects only in near-surface regions of the sample, changes can only be detected with insufficient certainty by transmission measurements.

For the adhesives investigated (3 M Scotch-Weld DP100 (2C epoxy) and Weicon Easy-Mix RK-7000 (2C acrylate)), the water-aged samples in the wet state are very well distinguishable from the unaged samples, but UV ageing does not cause any significant changes. Thus, general statements about the suitability of THz spectroscopy for determining the ageing condition of plastics and adhesives should only be uttered carefully.

3.2 *Examination of Layer Systems in Reflection*

In an industrial environment, transmission measurements are often not possible. Many parts are only accessible from one side or have metallic layers that inhibit the transmission of THz radiation. Therefore, a reflection approach is unavoidable in many cases. In the ideal case, all parts of a layered structure, resp. adhesive bond, can be characterized by a single THz experiment: At each interface, a part of the incoming THz pulse is reflected and another part is transmitted. For a structure as shown in Fig. 6, four reflected pulses are expected.

For reflection measurements it is also important that the difference in the refractive index at the interfaces is sufficiently large, otherwise the reflected portion of the THz

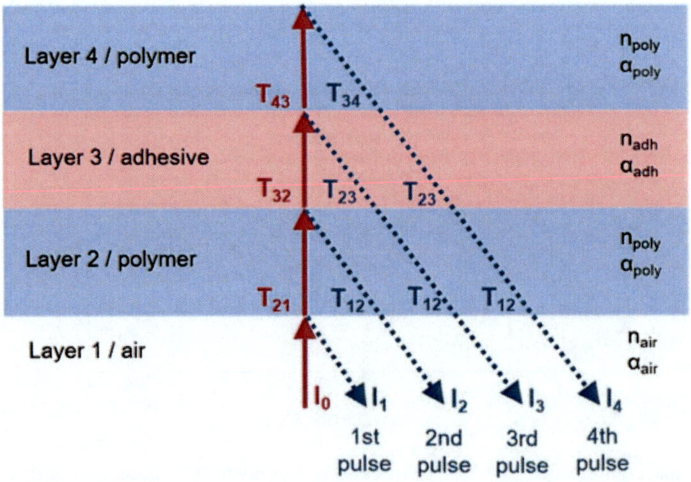

Fig. 6 Schematic representation of a THz reflection experiment on an adhesive bond

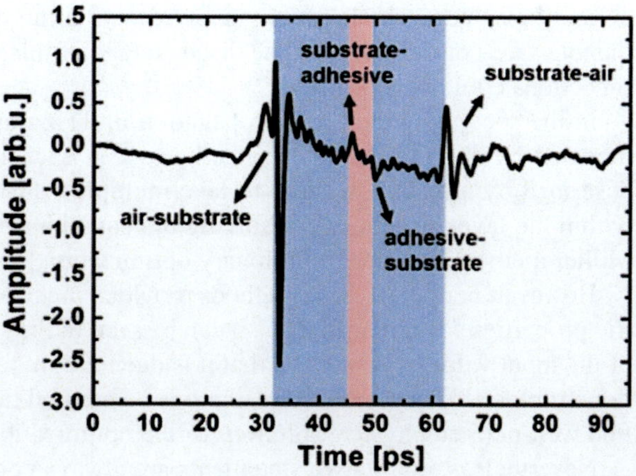

Fig. 7 THz measurement signal of a 3-layer system

pulse may be so small that it cannot be distinguished from the noise level of the experiment.

The reflected THz signal for the bond PA6/Weicon Easy-Mix RK-7000/PA6 (substrate thickness 1 mm, adhesive layer thickness 0.25 mm) is shown in Fig. 7. As can be clearly seen, the largest part is reflected at the air-substrate interface, another part at the substrate-adhesive interface, at the adhesive-substrate interface and finally at the substrate-air interface. For the exemplarily shown bond with an adhesive layer thickness of 0.25 mm, the distance between the two interface pulses substrate/adhesive is only in the range of 3 picoseconds (10^{-12} s). If the adhesive layer is even thinner, the reflected pulses may overlap. In this case, sparse deconvolution is necessary to separate the pulses from each other [30, 31].

In general, the estimation of dielectric parameters requires the fitting of a theoretical transfer function to a measured transfer function [32]. The measured transfer function is calculated by dividing the Fourier transform of the sample signal Y(ω) by the Fourier transform of the reference signal X(ω).

$$H_{\exp}(\omega) = \frac{Y(\omega)}{X(\omega)}$$

A reference is usually measured on a metal plate that is at the same distance from the emitter as the top surface of the sample. Moreover, even a slight deviation from this position leads to an error. An alternative approach is to place the sample on a metal plate. Then the reference signal is acquired on this plate and numerically shifted to match the top surface exactly. The first method was used here since the second approach is very sensitive to the layer thickness. Furthermore, it cannot be used in real applications, as the polymer often cannot be removed from the substrate in order to place them on metal for measurement.

In this work, we used two methods to fit the theoretical transfer function. In the first method, the transfer function was constructed by combining Fresnel's equations of each interface, which quantitatively describe the transmission and reflection at the interfaces of the sample. The disadvantage of this method is that not all internal reflections could be considered.

In the second method, the propagation of THz radiation inside the sample was modeled by the transfer-matrix method [33]: Each layer transition and layer are described by a matrix in order to take multiple reflections into account that occur within the layer structure. For solving the equations numerically, we tried both the Muller method [34] and evolutionary optimization [35].

However, neither of these methods provided precise results. The estimated dielectric properties scatter strongly as it can be seen in Fig. 8 and depend on the accuracy of the input values. We believe that it is due to high dimensionality of the optimization problem. To our knowledge, there is no method that can solve this problem and find well-defined physical solutions of the optimization problem.

Nevertheless, qualitative statements are always possible: Fig. 9 shows the peak-to-peak amplitude of the pulse reflected by the metal for a metal-acrylate adhesive-PA6 bond after water ageing. Ageing is dominated by the water absorption of PA6, resulting in a refractive index that increases with increasing immersion time (and thus a smaller proportion of transmitted radiation at the PA6-adhesive interface) and an increasing absorption coefficient (Fig. 3). For the acrylate adhesive Weicon Easy-Mix RK-7000, an increase in the absorption coefficient can also be determined in transmission measurements during long-term water immersion, which leads to a strong decrease of the signal of the reflected part of the metal in the adhesive bond. Only with prolonged ageing, the amplitude of the reflected portion increases again due to the decreasing absorption coefficient and refractive index of PA6.

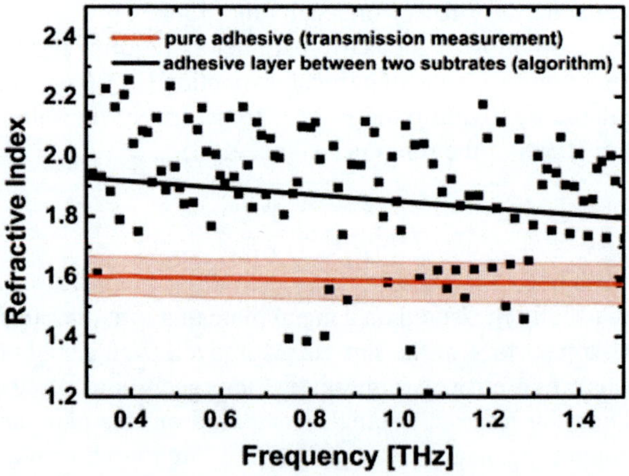

Fig. 8 Comparison of the frequency-dependent refractive index of the pure adhesive with that obtained for the adhesive in the layered composite by the Muller method

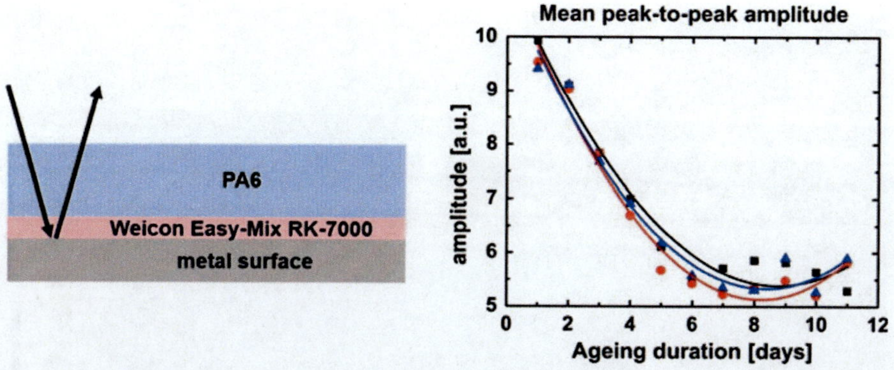

Fig. 9 Left: Investigated bonded joint which the THz pulse has to pass through; right: amplitude progression (peak-to-peak) of the signal reflected by the metal as a function of the ageing time (water immersion)

3.3 Imaging of Adhesive Seams

A THz spectrometer can also be operated in imaging mode [6, 36]. Using two mechanical linear stages, a pixel by pixel scan of the sample can be accomplished. For each pixel, the peak-to-peak amplitude at the corresponding position on the sample surface is then depicted by different colours (Fig. 10). THz imaging allows e.g. for the determination of birefringence [37] or water distribution in a sample [38].

For the imaging study presented herein, metal substrates (aluminum AlMg3 or steel DC04) were treated to obtain different surface conditions by means of cataplasm conditions for the generation of corrosion. Plastic substrates were bonded to these surfaces thus treated. The aim was to prove different substrate surface conditions (e.g. a surface conditions deviating from industrial standards or corroded areas) within the joints using THz sensor technology. An example of these investigations is shown in Fig. 10.

If the steel substrates show a high degree of corrosion, a strong change in the signal reflected at the metal interface can be observed in the THz image; the effect scales with the degree of corrosion or with the roughness profile determined by white light interferometry. R_{max} in the present case was 326.5 μm. Of particular interest for industrial applications is the fact that these changes are hardly influenced by an adhesive and subsequent plastic top layer, which makes it possible to detect rust infiltration on a bonded metal-plastic joint.

4 Summary

THz sensor technology can provide initial information about the fact that materials such as substrates and adhesives as well as interfaces in adhesive bonds are no

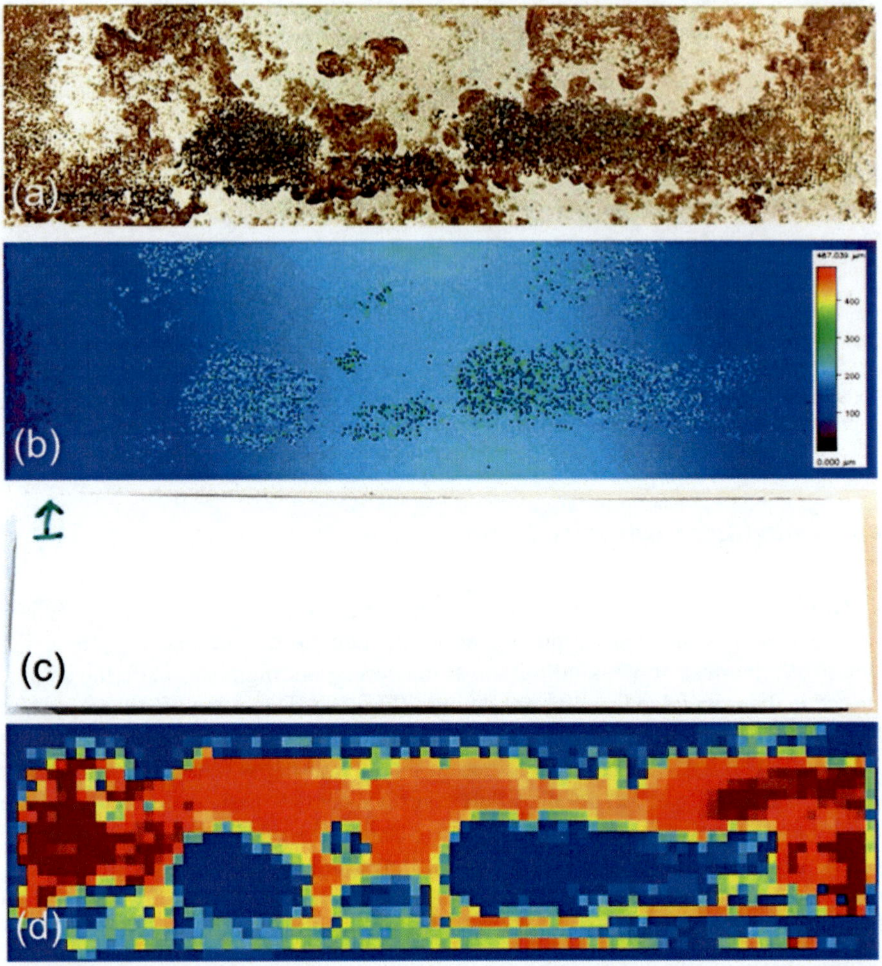

Fig. 10 Detection of rust between adhesive and metal substrate based on the THz radiation reflected at the interface, exemplarily for a compound steel DC04/adhesive 3 M DP100/polystyrene; with **a** optical image of corroded substrate before bonding, **b** roughness of corroded substrate before bonding, **c** optical image of the bonded specimen, **d** peak-to-peak amplitude of reflected THz pulse corresponding to the adhesive-metal interface of the bonded specimen (in dark red the highest amplitude, in blue the lowest)

longer in their unaged initial state. This can easily be achieved if interfaces in THz imaging are specifically examined for surface changes (e.g. rust) or layer structures for inclusions and foreign bodies, even if a top layer is present. Changes in pure substances can also be easily detected by transmission measurements, in particular if the substance embeds a high amount of water. However, the investigations in the underlying research project have clearly shown that it is difficult to interpret or even assign changes in the refractive indices or absorption coefficients to specific